HUMAN PHYSIOLOGY & HEALTH

MORTON JENKINS

Curriculum Manager in Science,
Coleg Glan Hafren, Cardiff

Hodder & Stoughton

A MEMBER OF THE HODDER HEADLINE GROUP

British Library Cataloguing in Publication Data

Jenkins, Morton, 1940–
   Human physiology and health
   1.Human physiology 2.Health
   I.Title
   612

   ISBN 0 340 658525

First published 1996

Impression number   10 9 8 7 6 5 4 3 2 1
Year   2000 1999 1998 1997 1996

Printed in Malaysia for Hodder & Stoughton Educational,
a division of Hodder Headline Plc, 338 Euston Road, London
NW1 3BH by Times Offset.

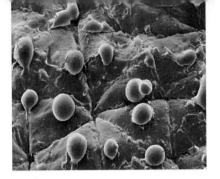

Contents

Contents

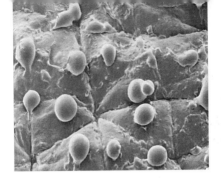

Acknowledgements

The author would like to thank the following people for their constructive criticism when reviewing the manuscript of this book:

Dr Sue Noake (Head Teacher, Lewis Girls School) and Mr Bill George (Deputy Head, Caerleon Comprehensive School).

The publishers would like to thank the following individuals, institutions and companies for permission to reproduce photographs in this book. Every effort has been made to trace and acknowledge ownership of copyright. The publishers will be glad to make suitable arrangments with any copyright holder whom it has not been possible to contact:

Allsport (59, 162 bottom); Heather Angel (143, right, 224 right); Biophoto Associates (77); Sir William Dunn, School of Pathology, University of Oxford (219 all right); Environmental Picture Library/Robert Brook (167 top right, 199 top, 221), /David Dennis (194 top), /Kiryu Hiroto (167 bottom left), /Steve Morgan (223), /David Sims (197); Hill and Knowlton (UK) Ltd (217 bottom right); Hulton Deutsch (219 top right); NHPA (38, 60 right), /Daniel Heuclin (164); Roddy Paine (3 left, 15, 158, 217 top right); Science Photo Library (143 left, 146 bottom, 203 left, 206 top); Science Photo Library/Michael Abbey (1, 2 top right), /A Barrington Brown (140 top), /Alex Bartel (37), /Biology Media (101 top, 167 top left), /Biophoto Associates (12 right, 13 right), /Bettina Cirone (90), /CNRI (2 bottom left, centre and bottom right), /Martin Dohrn (146 top right), /Ken Eward (206 bottom), /Cecil H Fox (97), /Simon Fraser, Royal Victoria Infirmary, Newcastle upon Type (69), /G F Gennaro (45 right, 102 bottom right), /Dr Martyn Gorman (101 bottom, 102 top right), /Petit Format/Nestlé (134 top right), /John Heseltine (217 bottom left), /Manfred Kage (134 bottom, 203 right), /Francis Leroy, Biocosmos (121), /Bill Longcore (167 top right, 176), /Dr P Marazzi (12 left, 222), /Jerry Mason (42 right), /Astrid and Hans-Frider Michler (96, 102 left), /Dr P Motta/Dept of Anatomy/'La Sapienza', Rome (45 left, 52 top), /Prof P M Motta and E Vizza (118), /National Library of Medicine (163 left), /David Nunuk (163 right), /Phillipe Plailly (140 bottom), /Phillipe Plailly/Eurelios (3 right), /Chris Priest and Mark Clarke (42 left), /Roger Ressmeyer, Starlight (60 left), /J C Revy (52 bottom), /Francoise Sauze (204), /Dr Gerald Schatten (2 top left), /Science Source (11), /St Bartholomew's Hospital (202), /James Steveson (207 left), /St Mary's Hospital Medical School (13 left, 219 bottom left) /Andrew Syred (32, 116), /Sheila Terry (70), /Alexander Tsiaras (226), /Dr E Walker (207 right), /M I Walker (136), /John Walsh (134 top left); Rex Features London (166, 224 left); Trip/Eye Ubiquitous (167 bottom right, 215), /J Okwesa (162 top), /H Rogers (194 bottom left, 195, 199 bottom), V Shuba (165), /A Tovy (156); ZEFA (146 top left).

Cover photo: false colour scanning electronmicrograph of a lymphocyte and a phagocyte, courtesy Don Fawcett /E Shelton/Science Photo Library.

The illustrations were drawn by Hardlines, Oxford.

The publishers would also like to thank the following for permission to reproduce copyright material:

Welsh Joint Education Committee, Northern Examinations and Assessment Board, Southern Examining Group, and Northern Ireland Schools Examinations and Assessment Council.

Preface

The author has always believed that the learning process should involve a mastery of certain fundamental concepts, rather than the learning of 'laundry lists' of names and facts for the sake of expanding a memory. A knowledge of such names and facts should be seen as a means to an end rather than an end in itself. From an initial understanding, the integration of structure and function provides a systematic approach to the study of physiology and health. The approach to learning by understanding and investigation has evolved successfully in thousands of secondary schools and is preserved in *Human Physiology and Health*.

The author has attempted to maintain the readability at a level suited to the average post-sixteen year old student. All new words and terms stand out in bold type. These words are defined the first time they are used and a glossary is provided as a further reading aid for developing a 'working' vocabulary.

The need to present concepts to students of varied abilities is recognised. In order to present the subject in a form that allows for comprehension at various levels, learning objectives for the student are indicated at the beginning of each chapter. These objectives serve to guide the student in studying the major concepts of the chapter. The questions for review at the end of each chapter reinforce the student's understanding of the material. Each theme ends with sample questions of the standard to be expected at GCSE level and are there to help those preparing for this standard of examination. The value of these questions is based on the premise that, for revision, 'Failure to prepare is preparation for failure'.

Morton Jenkins, 1996

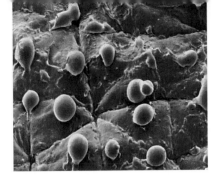

Introduction

A scientific method is a logical, orderly way of trying to solve a problem. It is this logic and order that makes scientific methods different from ordinary, hit-or-miss approaches. Remember, though, that scientific methods are not magic. Even the best planned investigations can fail to produce meaningful data. Yet failure itself may lead to final success. By careful study of each result, the scientist may find a new direction to take. Often this new direction leads to an even more important discovery than the one first expected.

Several methods are used in science, depending on the nature of the problem. Perhaps the most important is the **investigative method**.

Investigations

It is by investigating that new knowledge and new concepts are gained. There are many opportunities to investigate in your study of human physiology. The steps in this investigative method are logical and orderly. In fact, they are simply a system of common sense:

- Define the problem. You can't solve a problem unless you see that one exists. Science calls for the kind of mind that recognises problems and asks questions. For example, how does your intestine absorb digested food? Why does the pupil of your eye constrict in bright light? What controls your heartbeat?

Well planned investigations can answer each of these questions. However, in science, every answer raises new questions. Successful research leads to new research and new knowledge.

- Collect information on the problem. Scientists must build on the work of other scientists. Otherwise science could not advance beyond what one person could learn in one lifetime. Before beginning an investigation, the scientist studies all important informatiom that has to do with the problem. Often it turns out that someone has already answered many of the questions involved. For this reason, a library of scientific books and journals is an important part of investigating. This textbook will serve you as an aid to this background reading.
- Form an hypothesis. The available information may not fully solve the problem. The researcher must then begin to investigate. At this point, an hypothesis is needed. The hypothesis is a sort of working explanation or prediction.

It gives the investigator a point at which to aim. However, no matter how reasonable the hypothesis seems, it cannot be accepted until supported by a large number of tests. A research worker must be open-minded enough to change or drop the original hypothesis if the evidence does not support it.

- Test the hypothesis. The scientist must plan an investigtion that will either support or disprove the hypothesis. This means that the investigation must test only the idea or condition involved in the hypothesis. All other factors must be removed or otherwise accounted for. The one factor to be tested is called the single variable.

Often, a second investigation, called a **control**, is carried out along with the first. In the control, all factors except the one being tested are the same as in the first investigation. In this way, the control shows the importance of the missing variable.

- Observe and record data from the investigation. Everything should be recorded accurately. Correct, internationally standard units should be used. The record of the observations may include notes, drawings,

graphs, tables, etc. This information is called data and in modern research, data are often processed using computers or other information technology.

- Draw conclusions and evaluate. Data have value only when valid conclusions are drawn from them and validity can be assessed only when the investigation has been evaluated critically. Such conclusions and evaluations must be based entirely on observations made in the investigation. If other investigations continue to support the hypothesis, it may become a **theory**.

LIFE PROCESSES

The basis of life

Learning Objectives

By the end of this chapter, you should be able to:

- Describe the work of some early pioneers in the discovery of cells
- Explain the cell theory
- List the processes of a living cell
- Explain cell specialisation

- Explain the water relationships of cells in terms of diffusion and osmosis
- Explain why enzymes are important in cells
- Explain energy relationships of cells

The cell as the basic unit of life

All organisms, apart from bacteria and viruses, are made of one or more cells consisting of, at least, a nucleus, cytoplasm and a cell membrane (Figure 1.1.1). The more complex organisms may have thousands, millions, or even billions of cells. An organism's size depends on the number, not the size, of its cells. In general, elephant cells are no bigger than cells of an ant. An elephant simply has more cells.

A working knowledge of cells and their functions is fundamental to any study of human physiology and health. Indeed, almost every branch of biology deals, in some way, with cells.

The discovery of cells

More than three hundred years ago, in the 1660s, the British scientist, Robert Hooke, observed some thin slices of cork through a microscope (Figure 1.1.2). To his surprise, he said, the cork was a mass of cavities. Since each cavity was surrounded by walls like the cells in a honeycomb, Hooke called the structures he saw 'cells'.

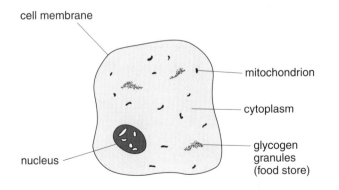

Figure 1.1.1 A typical animal cell

cell membrane
mitochondrion
cytoplasm
glycogen granules (food store)
nucleus

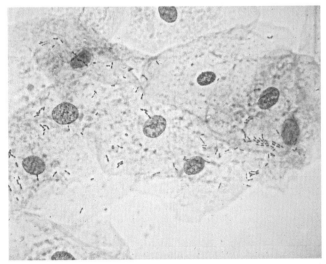

Micrograph of check cells ×500

Robert Hooke did not realise that the most important part of the cells that he described was missing. The empty cavities had once held living materials. Yet Hooke did not follow up his discovery by studying living cells. In fact, it was not until 1835 that another scientist discovered the living materials in cells. It was the French scientist, Dujardin, who viewed living cells with a microscope and found what we now call cytoplasm. Three years later, the German plant scientist, Matthias Schleiden, claimed that all plants are made of cells. The following year, Theodor Schwann, a German zoologist proposed that all animals are also made of cells. The discoveries of these contributed to the cell theory which states:

The cell is the unit of structure and function of all living things. Cells come from other cells by cell reproduction.

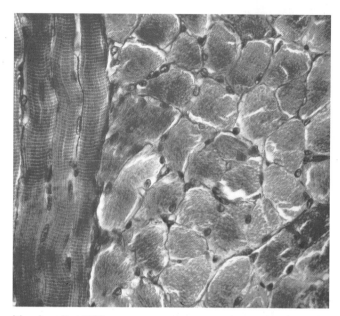

Muscle cells ×338

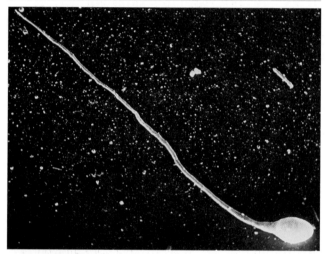

Sperm cell ×2662

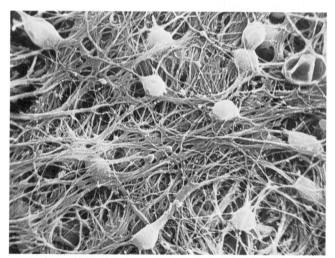

Nerve cells ×225

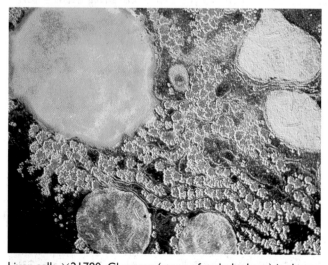

Liver cells ×21780. Glycogen (store of carbohydrate) is shown in red and fat in yellow

Skin cells ×63. You can also see a hair growing between two cells

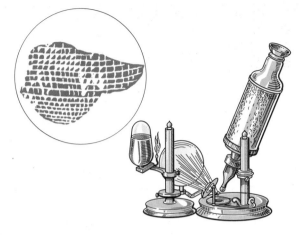

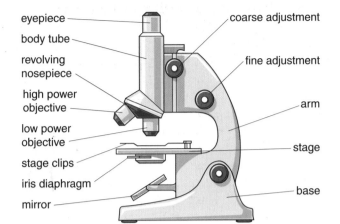

eyepiece
body tube
revolving nosepiece
high power objective
low power objective
stage clips
iris diaphragm
mirror
coarse adjustment
fine adjustment
arm
stage
base

Figure 1.1.2 Robert Hooke's cork section and his microscope

Figure 1.1.3 A light microscope

Since Dujardin's discovery, other biologists have added greatly to our knowledge of cells. However, they could not work beyond the limits of the light microscope which relies on lens systems and transmitted light (Figure 1.1.3). Then, in the mid-twentieth century, new tools and methods began to push back the limits to discovery. Most important was the electron microscope, which relies on a totally different system from the light microscope. It can produce magnifications of several orders of magnitude more than the magnification of a light microscope.

The most important parts of all cells

The **nucleus** is the control centre of all cell activity. Without a nucleus a cell will die.

The nucleus usually is a sphere or oval in shape, often lying near the centre of the cell. It is contained in a thin, double-layered membrane which has pores allowing certain

substances to pass between the nucleus and the rest of the cell.

Spread through the nucleus are fine strands of chromatin that make up the **chromosomes** which are so important in cell division (see page 133).

Cell chemicals outside the nucleus make up the **cytoplasm**. Under the light microscope, cytoplasm appears as a clear semi-liquid filling most of the cell. It often streams in the cell, carrying with it the nucleus and changing shape as it moves.

At its outer edge, the cytoplasm forms the **cell (plasma) membrane** which separates the cell from other cells and from surrounding fluids. Like the nuclear membrane, it is not a solid barrier. Molecules can pass through it in a controlled manner. It acts like a gate, allowing some molecules to pass and keeping others out (see page 5).

You may use a microscope like this in your investigations into cells

An electron miscroscope

Cells at work

Life depends greatly on the movement of certain materials into and out of the cell. Digested foods and oxygen are among the essential materials to be taken into cells. Whereas waste products such as carbon dioxide are removed. All materials must cross the cell membrane. Some molecules pass through freely, some less freely, and others not at all. Thus the membrane has a great effect on selecting materials for entry and exit. Several factors are important in determining this 'right of way'. Some of these are as follows:

- The size of the particles. Some large molecules cannot pass through the cell membrane. On the other hand, some large molecules pass through in greater numbers than much smaller ions (charged particles made of 'parts' of molecules).
- Whether or not the particles will dissolve in water. The liquids that bathe cells usually are solutions of molecules and ions in water. Substances which do not dissolve in water cannot pass through the cell membrane.
- Conditions inside or outside the cell. Conditions in the cell or in the cell's surroundings may affect the passage of particles. If there are more particles on the inside compared with the outside, they tend to move outwards.
- The structure of the cell membrane. The surface of the membrane seems to have many tiny pores. There must also be spaces between the molecules that make up the membrane. These spaces may be too small for large molecules to pass through, but large enough for smaller molecules.
- The rates and extent to which certain substances penetrate the cell membrane varies. If a substance passes through a membrane, we say that the membrane is **permeable** to that substance. If a membrane lets some substances pass but not others, we say that it is **selectively permeable** or **differentially permeable**. The cell membrane is selectively permeable.

Diffusion

To understand better how substances pass through cell membranes, we need to understand more about molecules and their motion. In any substance, the molecules are constantly moving. This motion results in **kinetic energy** within the molecules themselves and not from outside forces. Molecular motion is slight in solids, but much greater in liquids and gases. The movement of the molecules is completely random. That is, the molecules move in straight lines until they collide with other molecules. Then they bounce off and move in straight lines again until they collide with still other molecules. As you can imagine, this kind of movement results in a gradual spreading out of the molecules. In the end, they all will be spread evenly through any given space. This gradual, even spreading out of molecules is called **diffusion**.

As an example of diffusion, think of what happens if you open a bottle of perfume. As soon as you open the bottle, the molecules start diffusing into the air. Soon the people close to the perfume will smell it and, as diffusion continues, the smell becomes stronger. More and more molecules spread out at random among the molecules of gases in the air. Finally, a state of equilibrium is reached. The molecules, though still in motion, are spread evenly among the gas molecules in the air.

Actually, as the perfume diffused into the air, molecules of gases from the air also diffused into the perfume bottle. This brings us to a basic law of diffusion. According to this law, substances diffuse from areas of greater concentration to areas of lesser concentration. In this case, diffusion will continue until the concentration of gas molecules from the air and the perfume will be equal in all parts of the room.

Diffusion is affected by:

- Concentration – the greater the difference in

concentration between two substances, the more rapidly diffusion takes place.
- Temperature – the higher the temperature, the greater the speed of molecular motion.
- Pressure – the higher the pressure on particles, the greater the speed of diffusion from a region of high pressure to a region of low pressure.

Diffusion through membranes

How does a membrane affect the movement of molecules in diffusion? The answer depends on the nature of the membrane and the substance being diffused. You can demonstrate this with visking tubing and glucose solution. A sack is made out of the visking tubing and is filled with glucose solution (Figure 1.1.4).

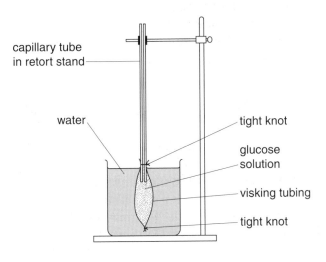

capillary tube in retort stand

water

tight knot

glucose solution

visking tubing

tight knot

Figure I.I.4 Experimental apparatus with glucose

The visking tubing sack containing the glucose solution is then placed in water. Since the visking tubing is permeable to both water and glucose molecules, two things will happen. Water molecules will diffuse into the glucose solution and glucose molecules will diffuse into the water. Finally, the concentration of glucose and water molecules will be the same on both sides of the visking tubing. This is a state of dynamic equilibrium because both types of molecule will be passing in both directions at

equal rates. The visking tubing had little effect on diffusion in this case.

What if the cell membrane were completely permeable? The answer is simple – the cell would die. True, molecules of water and other substances could enter the cell more easily, but at the same time, the cell's own molecules would diffuse out into the surroundings. Clearly, the cell membrane must only be selectively permeable if the cell is to survive.

You can demonstrate how a membrane can be selectively permeable by using larger molecules (Figure 1.1.5). Water molecules can pass through visking tubing but sucrose cannot. This is because the sucrose molecule is almost twice as large as the glucose molecule. If the apparatus is set up as before but having sucrose solution inside the visking tubing sack, water will pass into the sucrose solution but sucrose cannot pass into the water.

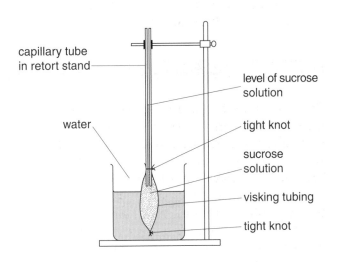

capillary tube in retort stand

water

level of sucrose solution

tight knot

sucrose solution

visking tubing

tight knot

Figure I.I.5 Experimental apparatus with sucrose

After about half an hour, the level of the solution in the tube will have risen, while the level of water in the beaker will have dropped. This is also an example of **osmosis**. There is diffusion of water through a selecively permeable membrane from an area of greater concentration of water molecules to an area of lesser concentration of water molecules.

So what does all this mean to our cells?

If the solution surrounding our red blood cells had less water than the cytoplasm of the red blood cells, water would pass out of the cells faster than it would enter. The cells would then shrink and die. If the opposite occurred and there was more water outside, then there would be a net gain of water and the cells would burst. Therefore, it is vital that the composition of our blood and other body fluids is regulated. This is an example of **homeostasis** or **regulation** of our internal environment (see page 10).

In many cases, certain particles pass in or out of cells against a **diffusion gradient**, i.e. against the direction of flow of particles. Sodium and potassium ions are examples of this and are actively transported in or out of cells under certain circumstances. Whereas diffusion is a passive process, active transport requires energy released by the cell to be used to 'pump' ions in one particular direction.

Enzymes – tools of a cellular factory

Cells are living chemical systems in which substances are constantly changing. Molecules react with other molecules. Large molecules are built up and broken down. What starts these changes? What controls them? What keeps one change from interfering with another? The answer is **enzymes**. They are proteins made by living organisms which alter the rate of chemical reactions. In fact, they can also be extracted from organisms and still retain their properties. We make use of this in **biotechnology** (see page 214) when we use enzymes in washing powders and certain food manufacturing processes.

It is important not to think of all enzymes as chemicals which always break large molecules into small ones like digestive enzymes. There are over one hundred thousand chemical reactions going on in your body at any given instant. Almost all of these are controlled by one or more enzymes, the vast majority of which have nothing whatsoever to do with digestion. Most are involved in releasing energy in the process of respiration.

All enzymes have certain properties in common:

- They are proteins.
- They are denatured when they are boiled. This means that their molecular structure is altered so that they can no longer work.
- They increase their rate of reaction with an increase in temperature up to a maximum.

This suggests that the optimum temperature for our enzymes will be body temperature.
- Their rate of reaction can be altered by changing the pH (degree of acidity or alkalinity). This means that each enzyme has an optimum pH in which it works fastest.
- They are specific in their action. This means that each one has a certain job. For example, enzymes used for energy release cannot be used for anything else, such as digestion or building up proteins.

Many possible models of enzyme action have been suggested. In one model, the enzyme is seen as being something like a piece in a jigsaw puzzle. Another compares the action to a lock and key fitting together (Figure 1.1.6).

At one place on the enzyme molecule is the **active site**. This site fits exactly with the molecule to be changed, which is called the **substrate**. A so-called, **enzyme-substrate complex** is produced. This complex reacts quickly, forming products from the substrate but not changing the enzyme.

Often, one chemical reaction in our bodies involves several different changes. This calls for a group or 'team' of enzymes. Such a group is called an **enzyme system**. Certain other molecules which are not proteins may take part in the enzyme activity. Such molecules are called **co-enzymes** and include many vitamins (see page 12).

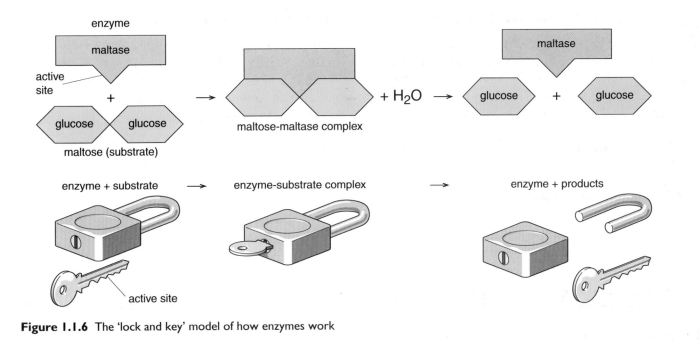

Figure 1.1.6 The 'lock and key' model of how enzymes work

Respiration – energy release in cells

Energy is the 'push' that makes things happen. It is the ability to do work and can be classified in two ways. If you wind up the spring motor of a clock, the tightly wound spring will have stored or **potential energy** in it. As the spring begins to unwind and turn the mechanism of the clock, the stored energy is changed into movement energy called **kinetic energy**.

When carbon atoms are joined to make large molecules of carbohydrates, the bonds which join the atoms are a form of potential energy. The bonds can be compared with wound-up springs.

A log, burning in a fireplace, is undergoing a chemical change in which the energy stored as chemical bonds is released. The carbohydrate in the log breaks down into carbon dioxide and water and the bond energy is rapidly converted into heat and light. We also give off carbon dioxide and water in a process which is similar to burning. Although the end products of the two processes are the same, cells 'burn' their fuel, glucose, in quite a different way. Instead of releasing energy rapidly, cells must save it to be used for many purposes. So chemical reactions in cells take place in small steps that release energy slowly. The overall reaction can be represented by the following equation:

$$C_6H_{12}O_6 + 6O_2 \rightarrow 6CO_2 + 6H_2O + Energy$$

glucose   oxygen   carbon   water
                        dioxide

However, the equation does not tell you much about the energy. It does not tell you the form that the energy takes and it does not show you that the energy is released in stages.

In fact, the energy is stored for a very short time before it is used, in the chemical bonds of a substance called **adenosine triphosphate (ATP)**. The ATP is a kind of chemical storage battery. In the ATP molecule, the three phosphate groups form a chain linked to the rest of the molecule.

A–P–P–P

The most important chemical fact about ATP is that the last phosphate group and bond can be relatively easily separated from the rest of the ATP molecule and can join onto other compounds. The loss of the phosphate group changes ATP into **adenosine diphosphate, ADP**, which can eventually be recycled to form ATP once more.

Energy from the chemical bond in ATP becomes available for the use of the cell in molecules and for all living processes. The release of energy from glucose in every living cell is called **respiration** and is a characteristic of all living organisms. The energy released allows us to demonstrate the other six characteristics, i.e. movement, growth, reproduction, sensitivity, excretion and nutrition.

Respiration is a controlled process and occurs in small steps, each helped by one or more respiratory enzymes. They allow the process to take place at normal body temperature. When oxygen is used in the process it is called **aerobic respiration** and is the normal form of respiration which occurs in our cells. However, it is possible for respiration to take place without oxygen. The process is called **anaerobic respiration** and it only releases a fraction of the energy that is released when oxygen is present. Glucose is again broken down with the use of respiratory enzymes but molecular oxygen is not used.

In some organisms, such as yeasts, ethanol and carbon dioxide are formed. Much use of this is made by biotechnology (see page 214) because it is the basis of baking and alcoholic fermentation.

$$C_6H_{12}O_6 \rightarrow 2C_2H_5OH + 2CO_2 + Energy$$
glucose ethanol carbon dioxide

In our muscle tissues, when no oxygen is available, glucose can be changed to lactic acid with the release of a little energy.

$$C_6H_{12}O_6 \rightarrow 2C_3H_6O_3 + Energy$$
glucose lactic acid

This is useful during physical exercise when muscle cells are sometimes starved of oxygen because the energy demand exceeds the supply of oxygen by the bloodstream. Under these circumstances, the body can still release energy for a short time, although there will be a build-up of lactic acid causing muscle fatigue. After such exertion, when at rest, we breathe more deeply and more rapidly. This is essential because a greater intake of oxygen is needed to pay off the **oxygen debt** incurred by the extra demand. The fitter we are, the more efficient is our circulation, and the less lactic acid is produced.

Summary

1 The cell is the basic unit of structure and function in living things.

2 Living things are organised according to different levels of structure. Some are made of a single cell, others are made up of many cells organised as tissues, organs, and organ systems.

3 The different kinds of cells in an organism support each other when they perform a particular function.

4 In order to remain in balance with its environment, the cell membrane controls movement of molecules in and out of the cell.

5 The cell membrane is selectively permeable so different particles pass through at different rates. Useful materials can pass through the membrane, but cell structures remain within the membrane.

6 Diffusion may cause some particles to move through the membrane. When water passes through it is said to pass through by osmosis.

7 Movement of particles by diffusion is called passive transport.

8 When cells absorb ions against a diffusion gradient, energy is used by the cell and so this is known as active transport.

9 Energy stored in glucose is released during respiration.

10 Glucose molecules are broken down and their chemical bond energy is released.

11 Some of this energy is given off as heat. Some is stored in ATP for use in cellular activities.

Questions for review

1 'In one sense Robert Hooke discovered cells, in another he did not.' Discuss this statement.

2 State the two principles of the cell theory.

3 List the levels of biological organisation from simple to complex.

4 Distinguish between permeable and selectively permeable membranes.

5 What is diffusion? What is the name for the balance that results from diffusion?

6 State the external factors that influence diffusion rates.

7 Define osmosis.

8 Distinguish between active and passive transport.

9 What is the biological importance of respiration?

10 Describe the two types of anaerobic respiration.

Applying principles and concepts

1 In what respect is the cell the basic unit of life?

2 Describe what might happen to a cell if the membrane were permeable to all molecules.

3 What factors determine whether or not a particle will pass through a plasma membrane?

4 Describe how cells derive their usable energy.

LIFE PROCESSES

LIFE PROCESSES

Nutrition

Learning Objectives

By the end of this chapter, you should be able to:

- Define food
- Explain the body's use of water
- List the food substances, their functions, and their sources
- Name and describe the organic nutrients

- Explain the importance of carbohydrates, fats and proteins
- Describe the structure and function of the parts of the digestive system

Food

What is food? What happens to a sandwich and a glass of milk that you have for lunch?

In a short while, the chemicals from them are in your body tissues. The process of preparing the foods to reach the body tissues occurs in a nine-metre tube called the **alimentary canal**.

Digestive enzymes break the bread, meat, milk and butter into smaller molecules of glucose, amino acids, glycerol and fatty acids. These can then pass into the bloodstream. Food is a collection of chemicals taken into an organism for growth, energy release, and for maintaining all the life processes. By this definition, water, minerals, and vitamins are food as well as carbohydrates, fats and proteins.

The many uses of water

Water is inorganic and gives no energy to the body tissues. However, it is essential for all forms of life. Humans can survive without any other food longer than without water.

If you weigh 50 kg, your body contains between 30 and 35 kg of water. Much of this water is in the protoplasm of your cells and in the spaces between cells. The fluid part of blood is up to 92% water. Water dissolves the food and waste that are transported to and from body tissues. It is also a solvent in the transport of dissolved foods from the alimentary canal to the blood. Waste materials, excreted from the skin and

kidneys are dissolved in water. Up to two and a half litres of water pass through the kidneys each day.

The loss of water by sweating also helps in regulating heat loss from our bodies. Evaporation, or the change from liquid to gas, requires heat. When your sweat evaporates from your skin, heat from your body is lost.

In these ways, we lose water from our bodies and this loss must be balanced by the amount of water we take in. We take in water in three ways (Figure 1.2.1):

1   From the food we eat.

2   As a by-product of chemical reactions in our cells.

3   From drinking all types of liquids.

What happens if you do not drink enough water? First, you lose water from between your cells. Then you lose it from the cells themselves. When this happens, the cytoplasm becomes affected. Finally the cells cannot function and they die as a result of dehydration.

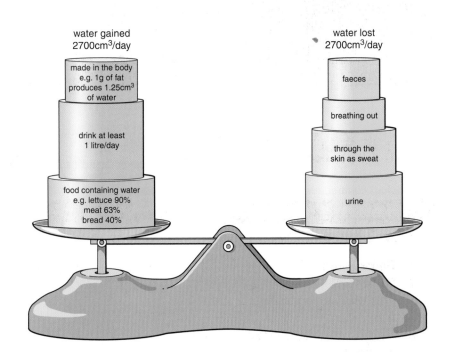

water gained
2700cm³/day

made in the body
e.g. 1g of fat
produces 1.25cm³
of water

drink at least
1 litre/day

food containing water
e.g. lettuce 90%
meat 63%
bread 40%

water lost
2700cm³/day

faeces

breathing out

through the
skin as sweat

urine

Figure 1.2.1 Water balance in humans

Mineral salts in the body

Food supplies us directly with salt (sodium chloride). There are also other mineral salts in our food. We lose salt when we sweat, and so people exposed to considerable heat for a long time must use extra salt in their food, or take salt tablets. We need calcium and phosphorus in greater amounts than other minerals. Calcium is needed for proper functioning of cell membranes and for blood clotting (see page 34). Together with magnesium, calcium is essential for nerve and muscle action. Phosphorus is needed for making the complex chemical, DNA, found in our chromosomes, (see page 140) and also in our temporary store of energy, ATP, formed during respiration (see page 7). Calcium phosphate is needed to form bones and teeth. Indeed, calcium and phosphorus make up about 5% of our tissue when they combine with the other elements in proteins. Milk is a good source of these two elements as well as wholegrain cereals, meat and fish.

Potassium compounds are needed for growth. Iron is needed for the formation of **haemoglobin** (see page 33) and is found most commonly in meat, green vegetables and a few types of fruits

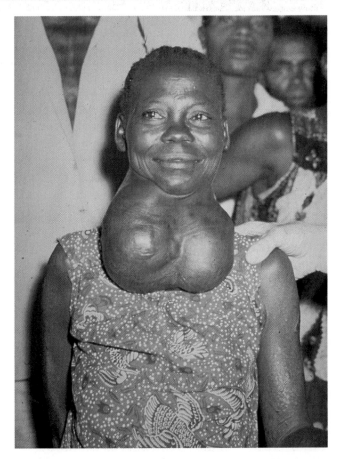

This woman has goitre caused by a lack of iodine in her diet

Table 1 Some essential minerals in food

Mineral	Essential for	Source
sodium compounds	blood and other body tissues	table salt, vegetables
calcium compounds	bone and teeth development nerve and muscle action blood clotting	milk, wholegrain cereals, meat, vegetables
phosphorus compounds	bone and teeth development DNA and ATP (see pages 7 and 140)	milk, wholegrain cereals, meat, vegetables
magnesium compounds	muscle and nerve action	vegetables
potassium compounds	blood cells and growth	vegetables
iron compounds	haemoglobin formation	liver, meat, spinach
iodine compounds	thyroxine formation (see page 88)	iodised table salt, sea foods

such as plums, raisins and prunes. Iodine salts are needed to form the hormone, thyroxine (see page 88). You can obtain iodine from iodised table salt or from any type of sea food.

Minerals are vital to the body in many ways but our bodies can use them only as compounds.

Eating chemically pure elements, such as sodium or chlorine, would kill you because they would harm your cells. As a compound, however, sodium chloride is essential.

Vitamins – essential for proper body function

In 1911, Dr Casimir Funk demonstrated that some substances in very small amounts in food seemed to be essential for normal body growth and activity. They were not ordinary nutrients and, without them, people suffer from deficiency diseases. At first, these essential substances were not identified and so they were given the letters, A, B, C, etc. instead of names. Today, the chemical nature of all **vitamins** is known and they all have chemical names. However, letters are still used for easy identification.

Vitamins are all organic substances, essential to life, but are not sources of energy. They help many of the enzymes which work inside your cells. Hundreds of years ago, when explorers went on long sea voyages, they ate only preserved foods. Many became ill with scurvy and died from bleeding of the gums and internal organs. It was then discovered that eating citrus fruits prevented scurvy and so ships began carrying barrels of limes on board.

What was in the citrus fruits that prevented scurvy? It was found to be vitamin C. Unlike some small mammals, like rats and hamsters, the human body cannot make this vitamin, so we must get it from our food.

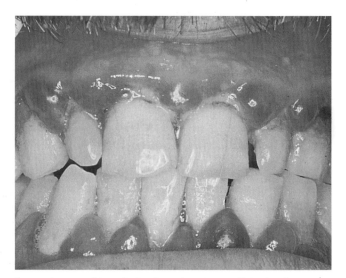

Scurvy is caused by a lack of regular vitamin C in the diet

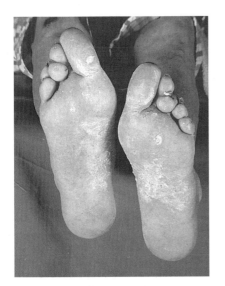

This person is suffering from pellagra caused by a lack of niacin in their diet

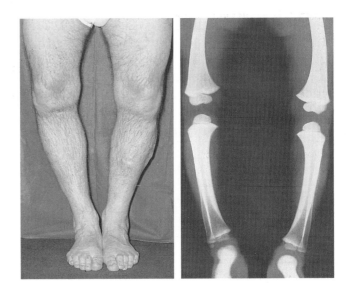

Lack of vitamin D causes rickets

Table 2 The functions and important sources of vitamins

Vitamin	Best sources	Essential for	Symptoms of deficiency
A (fat-soluble)	fish-liver oils, liver and kidney, green and yellow vegetables, butter, egg yolk	growth, eye-sight, skin and mucous-membranes	retarded growth, night blindness, viral infections
B_1 (water-soluble)	sea food, meat, soybeans, milk, wholegrain cereals, green vegetables	growth, energy-release, nerve and muscle action	retarded growth, nerve disorders, faulty digestion (beri-beri)
B_2 (water-soluble)	meat, soybeans, milk, green vegetables, eggs, yeast extract	growth, energy-release, healthy skin and mouth, eye function	retarded growth, dimness of vision, inflammation of the tongue
niacin (water-soluble)	meat, fish, wholegrain cereals, vegetables, potatoes, peanut butter	energy-release, functioning of intestine, functioning of nervous system	smoothness of tongue, skin disorders, digestive disorders, mental disorders (pellagra)
B_{12} (water-soluble)	liver, meat	preventing pernicious anaemia	reduction of red blood cells
C (water-soluble)	citrus fruit, vegetables, blackcurrants	growth, maintaining strength of blood vessels, development of gums and teeth	sore gums, bleeding around bones and from intestine (scurvy)
D (fat-soluble)	fish liver oil, liver, milk, eggs	growth, regulating calcium and phosphorus use for making bones and teeth	soft bones (rickets), dental decay
E (fat-soluble)	wheatgerm oil, vegetables, milk	reproduction	undetermined
K (fat-soluble)	vegetables, tomatoes, soybean oil	normal clotting of blood, normal liver functions	internal bleeding

We can store some vitamins which are soluble in fat, but those that are soluble in water cannot be stored and excess is excreted in our urine. Vitamin D can be made in our skin. You can get most vitamins in vitamin pills if your diet lacks any of them but if you eat a balanced diet, additional vitamins are unnecessary for the average person.

An investigation to find out how much vitamin C there is in fruit juice

Vitamin C, or ascorbic acid, is a powerful reducing agent and a dye, DCPIP, can be used to detect it. The vitamin causes the normally dark blue dye to be decolourised.

Method
Pour 1 cm³ DCPIP into a test tube. Take a 1 cm³ syringe full of sample material, e.g. lemon juice. Record the volume of sample material you need to add to the DCPIP to decolourise it. Repeat the test with 1 cm³ of fresh DCPIP using an ascorbic acid solution of known concentration. Record the volume of the second, standard solution needed to decolourise the DCPIP solution.

Sample results
Volume of lemon juice used to decolourise 1 cm³ of DCPIP solution = 0.5 cm³

Volume of ascorbic acid of known concentration used to decolourise 1 cm³ of DCPIP solution = 0.4 cm³

Conclusion
The lemon juice was 0.4/0.5 times as concentrated as the standard ascorbic acid solution.

If the standard ascorbic acid solution contained 1 mg ascorbic acid per cm³ water, then the lemon juice must contain 1 × 0.4/0.5 = 0.8 mg ascorbic acid per cm³ water.

Organic nutrients needed in bulk

We call carbohydrates, fats and proteins organic nutrients because they are formed by living cells and contain the element, carbon. **Carbohydrates** and **fats** supply most of our energy. The tissue-building value of foods can only be measured by observing growth of animals when they are fed, but energy values of food can be measured in **calories** or joules. One calorie is the amount of heat needed to raise the temperature of 1 gram of water 1 degree Celcius. One thousand calories equal one kilocalorie (Figure 1.2.2).

Today, heat energy is measured in the international units called joules.

> 1 kilocalorie = 4.2 kilojoules (kJ).

Table 3 Daily energy requirements of a variety of people

Person	Occupation/activity	Energy requirement (kJ)
newborn baby	sleeping, moving	1900
adult in bed	resting	7600
girl aged 8 years	very active in playing	8000
boy aged 8 years	very active in playing	8400
woman	office work	8800
man	office work	10500
pregnant woman	feeding her embryo	11500
girl aged 15 years	active in games/discos	11800
mother breast feeding	producing food for her baby	12600
man	moderately heavy work	14300
boy aged 15 years	active in games/discos	14700
man	heavy labouring work	18900

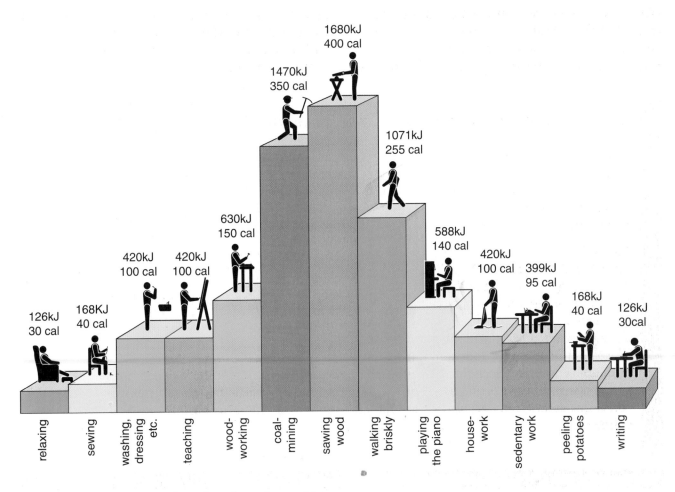

Figure 1.2.2 Our energy needs per hour for various activities

The chart shows energy needs per hour for various activities:

- relaxing — 126kJ / 30 cal
- sewing — 168KJ / 40 cal
- washing, dressing etc. — 420kJ / 100 cal
- teaching — 420kJ / 100 cal
- wood-working — 630kJ / 150 cal
- coal-mining — 1470kJ / 350 cal
- sawing wood — 1680kJ / 400 cal
- walking briskly — 1071kJ / 255 cal
- playing the piano — 588kJ / 140 cal
- house-work — 420kJ / 100 cal
- sedentary work — 399kJ / 95 cal
- peeling potatoes — 168kJ / 40 cal
- writing — 126kJ / 30cal

Why are carbohydrates important?

More than half of your food should be carbohydrate in a balanced diet. However, stored carbohydrate never makes up more than one percent of your body weight because carbohydrates are mainly fuel from which your cells release energy. The energy is released during respiration (see page 7).

You eat many different kinds of carbohydrates. Some are digested easily and then travel to tissues with little chemical change. Others have to be broken down before your tissues can use them. Some carbohydrates are not digested at all. We need these types in our diets for bulk, or roughage. All digestible carbohydrates reach body tissues as **glucose**.

There are many simple sugars in your food such as glucose and **fructose**. These are immediate sources of energy and require no chemical

change before blood can absorb them from the digestive organs.

Sucrose (cane sugar), **lactose** (milk sugar), and **maltose** (malt sugar) are larger molecules and

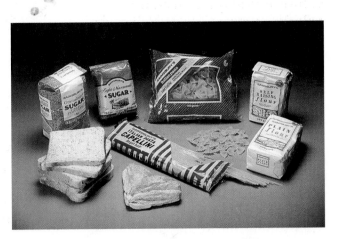

These foods are rich in carbohydrate

An investigation to measure the heat energy in a peanut

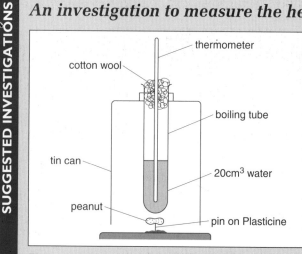

Figure 1.2.3 Apparatus for determining the energy in a peanut

Procedure

Weigh a large peanut and mount it on a pin as in the diagram. Place 20 cm³ water in the boiling tube. Note the temperature with a thermometer. Set light to the peanut with a Bunsen burner and place the boiling tube over it to catch as much of the heat as possible. (Use safety goggles.) Read the temperature as soon as the peanut has been completely burned. Record the temperature increase. Work

out the number of joules of heat the water has received as follows:

4.2 J raise 1 g water 1°C

Temperature increase = $Y°C$

Mass of peanut = X g

Mass of water = 20 g

Heat gained by water = $20 \text{ g} \times Y \times 4.2 \text{ J}$

Heat produced by

$$1 \text{ g of sample} = \frac{20 \times Yg \times 4.2 \text{ kJ per g}}{1000 \times X}$$

Note that the possible sources of error include:

- *heat loss around the sides of the boiling tube*

- *heat loss raising the temperature of the thermometer and the glass of the boiling tube*

- *incomplete burning of the peanut.*

An improvement would be to insulate the apparatus.

have to undergo digestion to break them into small molecules for absorption.

Starches make up a large part of the carbohydrates in most diets. There is a lot of starch in potatoes and cereals. Starches are made up of huge chains of glucose units. During digestion, starch is first changed to maltose, which is then broken down to glucose. Glucose is absorbed by the blood and carried to the body's tissues after first going to the liver. Much of the glucose that goes through the blood to the liver is turned temporarily into animal starch called **glycogen**. This can be turned back into glucose when the body needs it (see page 25).

Cellulose is a complex carbohydrate found in the walls of all plant cells. We cannot digest cellulose but it is important for the process of digestion. Cellulose stimulates muscle contractions throughout the alimentary canal and helps to pass the food along.

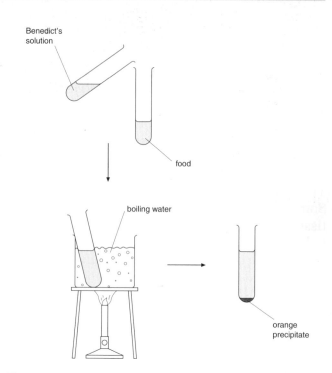

Figure 1.2.4 Testing for reducing sugar

Fats as an energy store

Fats and oils give more than twice as much energy as carbohydrates. Common sources of fats and oils include butter, cream, cheese, margarine, vegetable oils and meats.

During digestion, enzymes slowly break down fats. This happens in three stages. The result is one molecule of glycerol and three molecules of fatty acids from each fat molecule digested.

Excess carbohydrates are converted into fats and stored under the skin and around the kidneys. Too much fat is not good for you because it leads to obesity with all its problems of **cholesterol** build-up (see page 206). For this reason you should control how much carbohydrate and fat you eat. A balanced diet should consist of roughly, 60% carbohydrate, 20% fat, and 20% protein.

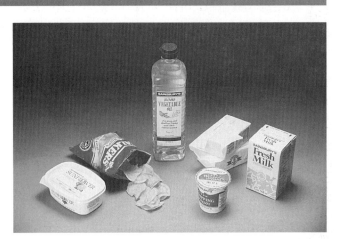

These foods contain a high fat content

Proteins and their uses

These foods are protein-rich

Proteins are complex organic molecules made of thousands of units called **amino acids**. They have to be broken down into these basic units during digestion so that they can be carried to tissues where they are built up again into human proteins. Proteins are used for growth and repair but they are also essential because all enzymes and hormones are proteins.

When we take in more protein than we can use we cannot store it. After being digested to amino acids, any that are there in excess are broken down into two parts in the liver during a process called **deamination**. One part (the amino part) contains nitrogen. It is changed to urea and transported by the blood to the kidneys. There it is excreted in the urine (see page 66). The other part, containing carbon, is changed to glucose and can be used by the body. Some of the best sources of protein include lean meat, eggs, milk, cheese, wholewheat and beans.

LIFE PROCESSES

SUGGESTED INVESTIGATIONS

Food tests

A test for starch
Method
Add a few drops of iodine dissolved in potassium iodide to a suspension of the food suspected to contain starch.

Expected observation
The suspension turns from its original colour to dark blue-black.

Conclusion
The food contains starch.

A test for a simple reducing sugar
Method
Add Benedict's solution to an equal volume of a solution suspected to contain a simple reducing sugar. Boil the mixture in a water bath.

Expected observation
The mixture changes colour from light blue, through green, then yellow, to an orange brown powder which settles as a precipitate.

Conclusion
The food contains a reducing sugar. (It reduced the copper salts in the Benedict's solution to copper oxide.)

A test for a soluble protein
Method
Add Biuret reagent to an equal volume of solution suspected to contain protein.

Expected observation
A violet colour is seen in the mixture.

Conclusion
The solution contains protein.

A test for fat
Method
Rub a sample of the suspected fat on a piece of paper.

Expected observation
A translucent patch is seen on the paper which remains when it is dried.

Conclusion
The food tested contains fat.

The stages of digestion

Why can't your body's tissues use most foods in the form in which you eat them? There are two reasons. First, many foods will not dissolve in water. This means that they could not get through cell membranes even if they could reach them. Second, the foods you eat are chemically complex. Cells cannot use them either to release energy or to make proteins. **Digestion** solves both these problems. During digestion, complex foods are broken down into small water soluble molecules. These molecules can be absorbed and used by your cells.

Digestion occurs in two stages. The first is mechanical. You chew the food, then the muscular movement of the wall of the digestive system churns and mixes it with various juices. All of this aids the second stage of digestion which is chemical. In this stage, digestive enzymes, secreted by the digestive glands, complete the job.

The digestive system includes the organs that form the **alimentary canal** (Figure 1.2.5). Other organs in the digestive system do not actually receive undigested food. Instead, they deliver secretions into the alimentary canal through ducts. Ducts are tubes that go from certain glands into the organs where food is being digested.

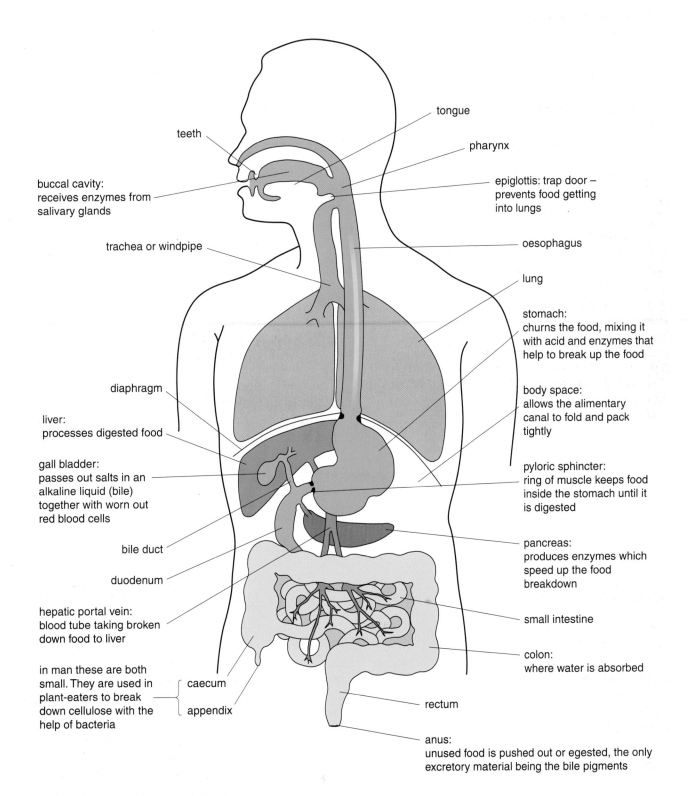

teeth

tongue

pharynx

buccal cavity:
receives enzymes from
salivary glands

epiglottis: trap door –
prevents food getting
into lungs

trachea or windpipe

oesophagus

lung

stomach:
churns the food, mixing it
with acid and enzymes that
help to break up the food

diaphragm

body space:
allows the alimentary
canal to fold and pack
tightly

liver:
processes digested food

gall bladder:
passes out salts in an
alkaline liquid (bile)
together with worn out
red blood cells

pyloric sphincter:
ring of muscle keeps food
inside the stomach until it
is digested

bile duct

pancreas:
produces enzymes which
speed up the food
breakdown

duodenum

hepatic portal vein:
blood tube taking broken
down food to liver

small intestine

in man these are both
small. They are used in
plant-eaters to break
down cellulose with the
help of bacteria

caecum

colon:
where water is absorbed

appendix

rectum

anus:
unused food is pushed out or egested, the only
excretory material being the bile pigments

Figure 1.2.5 The alimentary canal

parotid gland

duct of the parotid gland (to cheek)

duct of the sublingual gland

sublingual gland

neck vertebrae

submaxillary gland

oesophagus

nasal cavity

mouth cavity

tongue

teeth

lower lip

duct of the submaxillary gland

lower jaw bone

larynx

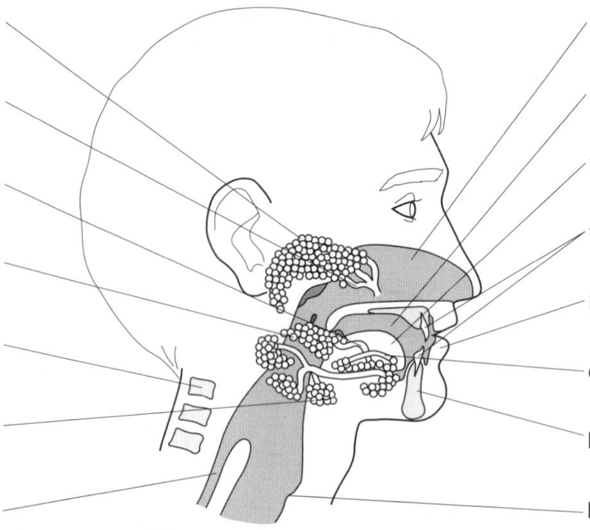

Figure 1.2.6 The mouth and salivary glands

The mouth

Your mouth's chief job is to prepare food for digestion. Salivary glands secrete saliva into the mouth through ducts located opposite your upper back (molar) teeth and also in the floor of the mouth, under the tongue (Figure 1.2.6).

When your mouth 'waters', these glands are secreting saliva. This happens when you taste food or simply when you smell or see it. Even thinking of food, when you are hungry, can start the secretion of saliva.

The types of teeth

We have two sets of teeth during our lives. These are the milk teeth and the permanent teeth. Milk teeth appear first and are not so strong as our final permanent teeth. They also lack wisdom teeth (molars). The permanent teeth are arranged the same in the top and bottom of the jaws (Figure 1.2.7).

The two flat front teeth are called **incisors**. They have sharp edges for cutting food. Next to the incisors, at the corner of your lips on either

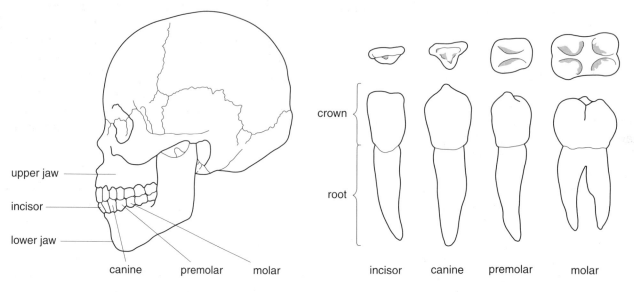

upper jaw

incisor

lower jaw

canine   premolar   molar

crown

root

incisor   canine   premolar   molar

Figure 1.2.7 The skull and teeth

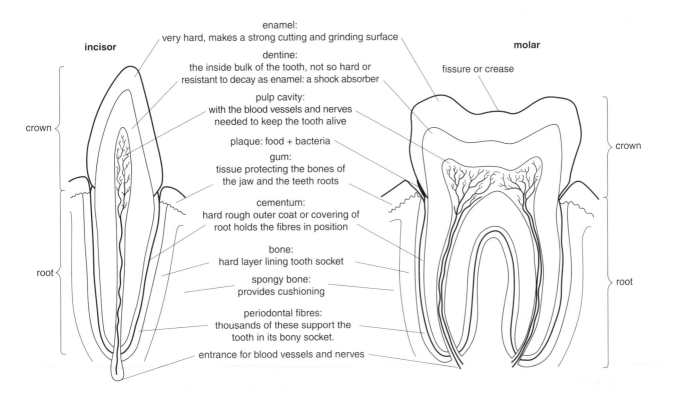

Figure 1.2.8 The structure of an incisor and a molar

side, is a large, cone-shaped tooth called the **canine**. Behind the canine tooth are the two **premolars**. Next are the **molars**. You have three of these on either side if you have your complete set. However, your molars may not appear until you are in your twenties. Premolars and molars have flat surfaces which are good for grinding and crushing. Many jaws are too small too hold the third molars, or wisdom teeth. In these jaws, wisdom teeth often grow crooked or remain impacted in the gums and may have to be removed.

The structure of a tooth

A tooth has three general areas (Figure 1.2.8). The part above the gum is called the **crown**. A narrow part at the gum line is called the **neck**. The **root** is the part beneath the surface. The root is held in a socket in the jaw bone. A fibrous **periodontal membrane** anchors it firmly in the jaw socket. Different kinds of teeth have different shaped roots. Some are long and single. Some have two, three, or four projections. The covering of the root is called **cementum**. It holds the tooth firmly in place. The crown has a hard white covering called **enamel**.

If you cut a tooth lengthwise, you can see the dentine beneath the enamel and cementum. Dentine is softer than these protective layers. It forms the bulk of the tooth. The **pulp cavity** lies inside the dentine area. The pulp cavity contains blood vessels and nerve fibres.

The oesophagus and stomach

Food that we swallow passes into the **oesophagus**. It is a tube about 30 cm long connecting the mouth to the stomach. Layers of muscle line the wall of the oesophagus helping the food to move to the stomach. One layer is circular and squeezes inwards. The other layer runs lengthwise. It contracts in a wave that travels downwards. The food is pushed ahead of the contraction in a process called **peristalsis** (Figure 1.2.9).

The **stomach** is in the upper part of the **abdomen**, just below the **diaphragm** (Figure 1.2.10). The walls of the stomach have three layers of muscle. One layer is circular; one runs the length of the stomach; and the third is arranged at an angle. The muscle fibres of these layers contract in different directions. This action causes the stomach to twist, squeeze, and churn.

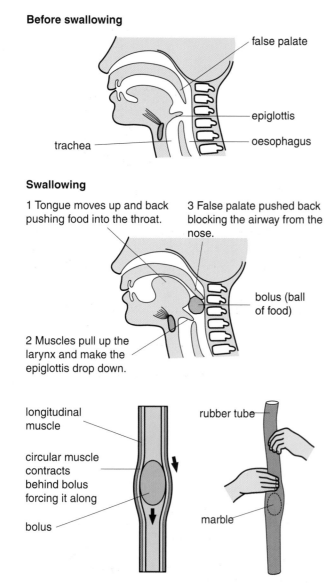

Before swallowing

- false palate
- epiglottis
- oesophagus
- trachea

Swallowing

1 Tongue moves up and back pushing food into the throat.

3 False palate pushed back blocking the airway from the nose.

2 Muscles pull up the larynx and make the epiglottis drop down.

bolus (ball of food)

- longitudinal muscle
- circular muscle contracts behind bolus forcing it along
- bolus
- rubber tube
- marble

Figure 1.2.9 Swallowing and peristalsis

The stomach lining is a thick, wrinkled membrane (Figure 1.2.11). It contains many **gastric glands**. Each gland is a tiny tube that opens into the stomach. Cells that secrete digestive juice line the tube. There are three kinds of gastric glands. One kind secretes digestive enzymes; another secretes hydrochloric acid; and a third secretes mucus. The mixture of these secretions is called **gastric juice**.

Food usually stays in the stomach for two to three hours. Rhythmic muscular contractions churn the food back and forth in a circular path. This churning separates food particles and mixes them thoroughly with the gastric juice. When the stomach finishes digesting the food, a ring of muscle, called the **pyloric sphincter**, relaxes, allowing food into the **small intestine**. When the stomach is finally empty, it rests for a while. After several hours without food, the stomach starts contracting again, making you feel hungry.

The small intestine

Food moves from the stomach to the small intestine. This is a tube about 3 cm in diameter and 7 m. long. The first 25 cm are called the **duodenum** which curves upward, then back to the right, beneath the liver. Beyond the duodenum is the much longer **ileum**. It is about 5 m. long and coils through the abdominal cavity. The end of the ileum joins the **large intestine**.

The mucous lining of the small intestine has many tiny **intestinal glands** which secrete **intestinal fluid** into the small intestine. This fluid contains enzymes used in digestion.

The liver – your largest gland

The **liver** weighs about one and a half kilos. It is a dark chocolate colour and lies in the upper right area of your abdomen (Figure 1.2.12). The liver secretes **bile**, a brownish-green fluid which passes through a series of **bile ducts**. The secreted bile travels to the **gall bladder** where it is stored. Here bile is concentrated as part of its water is removed. The larger common bile duct carries bile from the gall bladder to the duodenum. Sometimes this bile duct becomes clogged by **gallstones**. This makes the bile enter the bloodstream and cause jaundice which turns the eyes and skin a yellowish colour.

The pancreas and pancreatic juice

The **pancreas** looks similar to a salivary gland in that it is white with many lobes. It lies behind the stomach against the back wall of the abdominal cavity (Figure 1.2.13). The pancreas performs two very different functions. It passes a digestive secretion called **pancreatic juice** into the small intestine through the **pancreatic duct**. This duct leads to a common opening with the bile duct in the wall of the duodenum. The pancreas also produces the hormones, **insulin** and **glucagon** (see page 92).

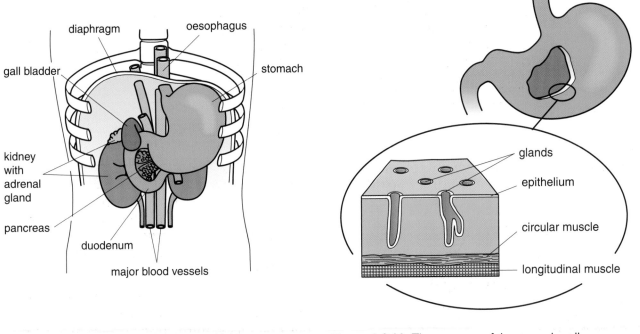

Figure 1.2.10 The position of the stomach

Figure 1.2.11 The structure of the stomach wall

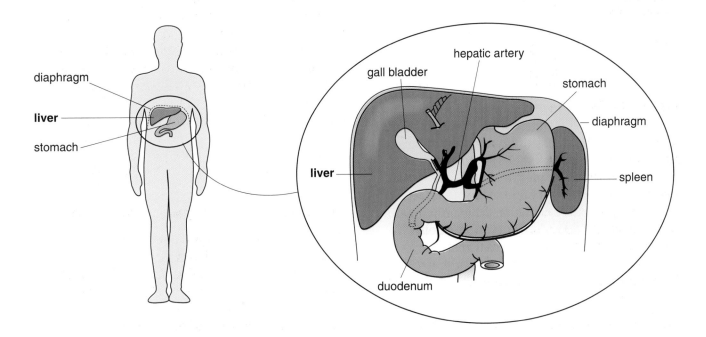

Figure 1.2.12 The position of the liver

The large intestine

The lower end of the small intestine connects with the colon in the lower right part of the abdominal cavity. The **caecum** is located where they both meet. Leading from the caecum is the finger-like **appendix**. This is the part which may become infected and cause appendicitis. The colon is about seven cm in diameter and one and a half metres long. It forms an upside-down, U-shaped structure in the abdominal cavity. At the end of the colon is the **rectum** leading to the **anal opening**. A ring-like sphincter muscle in the lower end of the rectum controls the elimination of intestinal waste called **faeces**.

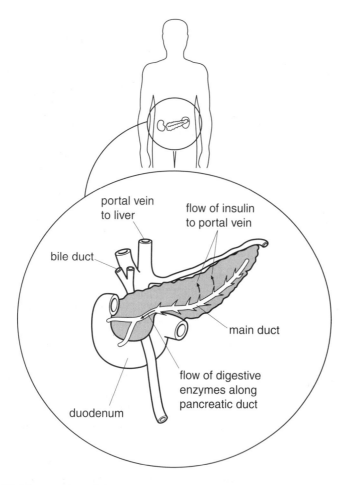

portal vein to liver

flow of insulin to portal vein

bile duct

main duct

flow of digestive enzymes along pancreatic duct

duodenum

Figure 1.2.13 The position of the pancreas

The chemical stages of digestion

As foods move along the alimentary canal, they go through a series of chemical changes. This is a step-by-step process of breakdown of large insoluble molecules to smaller soluble molecules. At each stage, a specific **digestive enzyme** is used. Each enzyme splits specific kinds of molecules. The digestive enzymes in the alimentary canal act outside the cells and therfore are responsible for **extracellular digestion**.

Digestion in the mouth

Chemical digestion begins in the mouth. Here, a **salivary enzyme** begins the digestion of starch (Figure 1.2.14). Saliva is about 95% water. It also contains mineral salts to regulate pH, lubricating mucus, and the enzyme, amylase (a type of carbohydrase). Amylase changes starch into maltose. Starchy foods, like potatoes, should be cooked before eating. This bursts their cellulose cell walls and allows the amylase to reach the starch grains. Food is only in the mouth for a short time so starch digestion is seldom finished when the food is swallowed.

Gastric juice at work

The main enzyme in gastric juice is **pepsin**. This enzyme works on protein, splitting the complex molecules into simpler groups of amino acids called **peptones** and **proteoses**. This splitting is the first in a series of chemical changes involved in protein digestion.

Hydrochloric acid is produced in the stomach. It helps pepsin work and also kills many

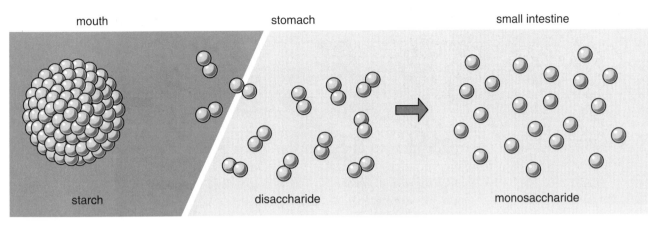

Figure 1.2.14 The breakdown of carbohydrate

bacteria which may enter the stomach with food.

What does the food that passes from the stomach to the small intestine contain? It contains: i) fats, unchanged; ii) the sugars, sucrose, glucose and lactose unchanged; iii) maltose sugar, formed by amylase acting on starch; iv) any starches not yet changed by amylase; v) peptones and proteoses formed by pepsin acting on protein; and vi) any proteins not yet acted on.

The liver and bile

The liver has many functions. Here are some of the more important ones:

- It uses glucose from the blood and changes it to glycogen which it stores as a reserve of carbohydrate.

- In the liver excess amino acids which we cannot use for growth, etc. are changed to urea. So the liver is also an organ of excretion.

- The liver secretes bile which has many important characteristics.
  i) In part, it is a waste substance because it contains materials from used, dead red blood cells.
  ii) Bile is NOT a digestive enzyme but it helps the process of fat digestion because it breaks globules of fat into an emulsion. Lipase from the pancreas can act on fats more easily in this form because a larger surface area of fat is exposed to the lipase (Figure 1.2.15).
  iii) It neutralises acid from the stomach so that pancreatic enzymes can work.

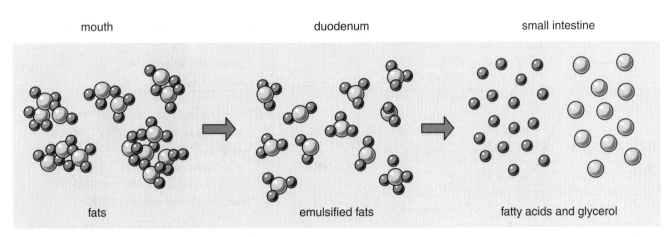

Figure 1.2.15 The breakdown of fat

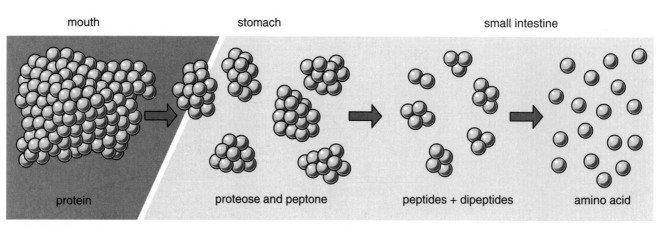

mouth   stomach   small intestine

protein   proteose and peptone   peptides + dipeptides   amino acid

Figure 1.2.16 The breakdown of protein

- The liver releases much heat because of all the chemical reactions taking place there. In this way it helps to maintain a constant temperature because the heat is carried through the blood in the same way as heat is carried in the water of a central heating system.
- It stores many useful materials such as iron (used to make red blood cells in the bone marrow) and vitamins A and D.

The pancreas and digestion

Pancreatic juice has enzymes to act on all three of the main classes of food. The enzymes are as follows:

- **Trypsin** which continues the breakdown of proteins that began in the stomach. It changes peptones and proteoses into simpler molecules called **dipeptides** (Figure 1.2.16).
- **Pancreatic amylase** which changes any remaining starch to maltose sugar.
- **Lipase** which splits fats into fatty acids and glycerol.

Table 4 A summary of digestion

Place of digestion	Glands	Secretion	Enzymes	Digestive action
mouth	salivary	saliva	amylase	starch to maltose
stomach	gastric	gastric juice hydrochloric acid	pepsin	proteins to peptones and proteoses activates pepsin kills bacteria
small intestine	liver pancreas	bile pancreatic juice	trypsin amylase lipase	emulsifies fats peptones and proteoses to peptides starch to maltose fats to fatty acids and glycerol
	intestinal	intestinal juice	erepsin maltase lactase sucrase lipase	peptides to amino acids maltose to glucose lactose to glucose and galactose sucrose to glucose and fructose fats to fatty acids and glycerol

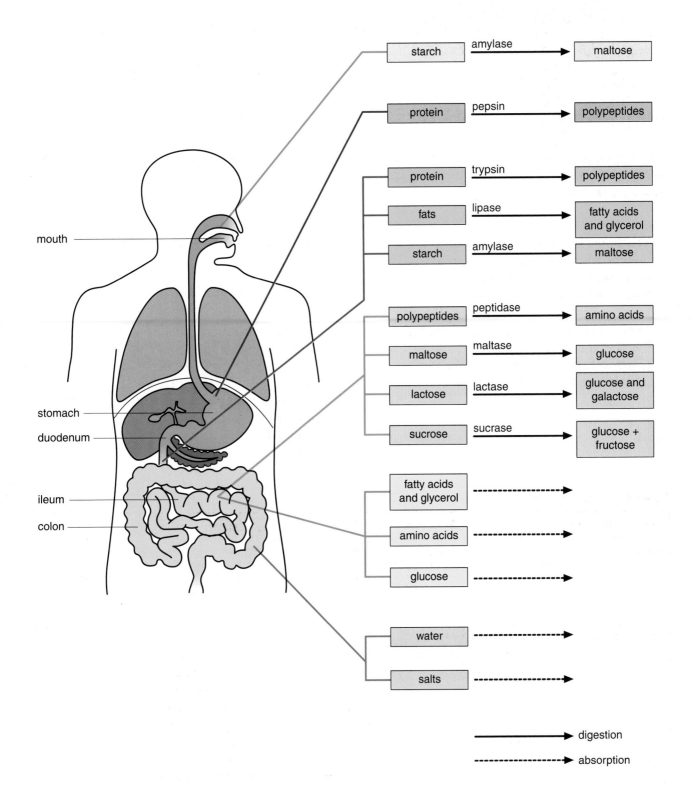

Figure 1.2.17 Enzymes in the alimentary canal

The small intestine and digestion

Here, digestion is completed and to accomplish this, at least five enzymes are involved:

- Erepsin completes the digestion of protein to amino acids.
- Lipase completes the digestion of fats to glycerol and fatty acids.
- Maltase breaks maltose into glucose.

- Sucrase breaks sucrose (sugar from sugar cane and sugar beet) into glucose and fructose.
- Lactase breaks lactose (milk sugar) into glucose and galactose.

The products of digestion are now soluble and are ready to leave the alimentary canal by absorption so that they can be carried to where they are needed in the body (Figure 1.2.20).

An investigation into the action of amylase on starch

Procedure
Label three test tubes 1, 2 and 3. Add 1% starch suspension to each. Boil 2 cm³ of 1% amylase in a water bath for 1 minute and add it to test tube 1. Add 2 cm³ of distilled water to test tube 2. Add 2 cm³ of unboiled 1% amylase to test tube 3. Use a pipette to withdraw one drop of mixture from each of the test tubes 1, 2, and 3.

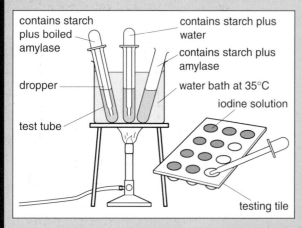

Figure I.2.18 Experimental apparatus

Add the drops to iodine in potassium iodide solution on a white tile as shown. Record your observations in a table. Place the test tubes in a water bath at 35°C.

Test one drop of each mixture at two minute intervals up to twenty minutes. Record your observations. Plan and carry out a method of detecting the presence of the substance produced as a result of the reaction in test tube 3. Plan and carry out investigations at varying temperatures.

An investigation of the action of pepsin on egg

Procedure
Take the white (albumen) of an egg and draw it into five pieces of capillary tubing, each 2 cm long. Put the capillary tubes into a beaker of boiling water and leave them for two minutes. The capillary tubes will then contain hard boiled egg white. Measure the lengths of the egg white in each tube. Place the capillary tubes in test tubes as shown in the diagram.

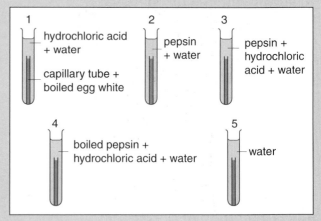

Figure I.2.19

Place these test tubes in a water bath at 35°C for 30 minutes. Then measure the lengths of egg white in each tube again. Record your results. State your conclusion and evaluate the procedure.

section of small intestine

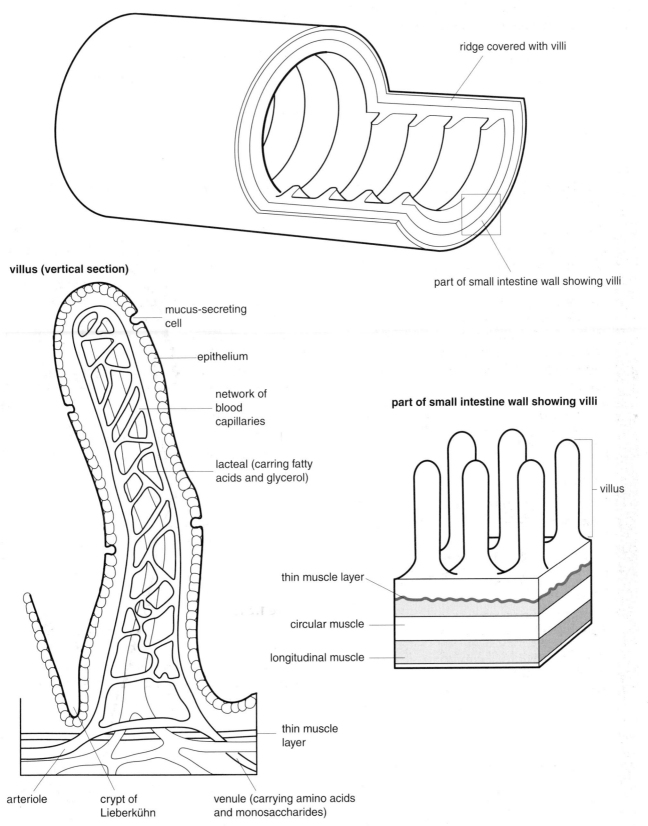

ridge covered with villi

part of small intestine wall showing villi

villus (vertical section)

mucus-secreting cell

epithelium

network of blood capillaries

lacteal (carring fatty acids and glycerol)

part of small intestine wall showing villi

villus

thin muscle layer

circular muscle

longitudinal muscle

thin muscle layer

arteriole

crypt of Lieberkühn

venule (carrying amino acids and monosaccharides)

Figure 1.2.20 The structure of the small intestine

Absorption

The small intestine has many finger-shaped projections in its irregular lining. These projections are called villi. There are so many of them that they give the intestinal wall a velvety appearance and present an enormous total surface area in contact with the digested food for absorption into the transport system of the body. The small intestine is remarkably well adapted for its two important functions. These are:

- The completion of digestion. It has intestinal glands to secrete intestinal juice.
- Absorption. It has
  i) Millions of villi to increase its surface area.
  ii) A rich blood capillary supply to carry away glucose and amino acids. The capillaries join up to form the hepatic portal vein which carries blood directly to the liver.

iii) Lacteals to carry away glycerol and fatty acids.
iv) A very thin lining so that small molecules can pass across to the capillaries or lacteals.
v) A moist lining because the food molecules will only be absorbed when in solution.

Water absorption in the large intestine

The large intestine receives watery masses of undigestible material from the small intestine. The water is valuable and is recycled by being absorbed into the rich blood supply of the colon and the rectum. The waste material, called **faeces**, is largely made of undigested cellulose fibres and dead bacteria. It is eliminated through the anus.

An investigation into absorption in the alimentary canal

Procedure
Set up the apparatus as shown in the diagram.

Test the distilled water with Benedict's reagent (see page 16) and with iodine in potassium iodide (see page 19) as soon as you have set up the apparatus. Repeat the tests at two minute intervals by taking further samples from the boiling tube. Record your observations in the form of a table. Explain the observations you have made. What part of the alimentary canal does the visking tubing represent? What does the distilled water represent? Explain how the observations illustrate the importance of digestive enzymes.

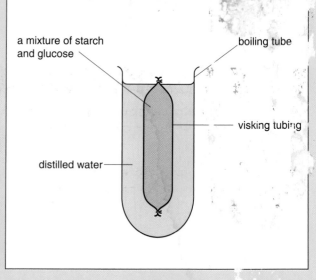

a mixture of starch and glucose

boiling tube

visking tubing

distilled water

Figure 1.2.21 Experimental apparatus

Summary

1 Our bodies need carbohydrates, proteins, fats, water, minerals, and vitamins.

2 The digestive system is a tube divided into various regions. This tube is called the alimentary canal. Each region is specialised as an organ. Each organ is adapted for performing certain stages of the digestive process.

3 Both our teeth and muscles of the alimentary canal break down mechanically the food we eat.

4 Many glands secrete enzymes into the alimentary canal. These break down food chemically.

5 Digestion must take place before food can be absorbed and used by our cells.

Questions for review

1 What are the functions of the main classes of foods?

2 Explain why the body must have water.

3 In what two general ways must food be changed during digestion?

4 List, in order, the divisions of the alimentary canal. What digestive processes occur in each?

5 Name the parts of a tooth.

6 Why is it especially important that you chew bread and potatoes thoroughly?

7 Suppose that you had a glass of milk and a sandwich made of bread, butter, and chicken. Describe what would happen to each of these foods as digestion occured.

8 Name an important function of the large intestine.

Applying principles and concepts

1 Explain how a vitamin deficiency is possible even if an adequate amount of all the vitamins is taken daily.

2 Why is it an advantage to have acid conditions in the stomach and alkaline conditions in the small intestine?

3 What is the advantage of peristalsis?

4 Why is it easier to digest sour milk than fresh milk?

LIFE PROCESSES

LIFE PROCESSES

Circulation

Learning Objectives

By the end of this chapter, you should be able to:

- Define blood
- List the plasma components and the cells
- Describe the function of each part of the blood
- Explain how the blood clots
- Understand the ABO blood grouping system

- Discuss how the Rh factor may affect babies
- Trace the circulation of blood through the heart
- Name and describe the types of blood vessels
- Define lymph

A fluid tissue

The fluid that carries all of your body's needs is called **blood**. An average person has about six litres of blood which makes up about 9% of your body mass. Blood is a type of **connective tissue** (see page 96) which is made of two different parts. There is a non-living fluid part, called plasma, and there are living **blood cells** floating in the plasma (Figure 1.3.1).

Plasma is a sticky, straw-coloured liquid, 90% of which is water. Proteins in the plasma make it sticky and one of these is called **fibrinogen**, used in blood clotting. **Serum albumin** is another plasma protein necessary for the absorption of materials into blood. The third is **serum globulin** which gives rise to the **antibodies** that lead to **immunity** to various

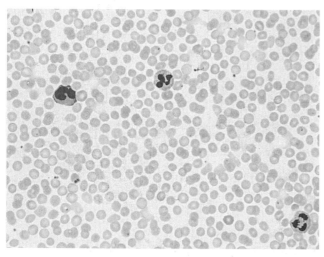

A photomicrograph of blood cells ×968

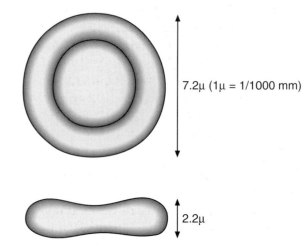

7.2μ (1μ = 1/1000 mm)

2.2μ

Figure 1.3.1 Red blood cells

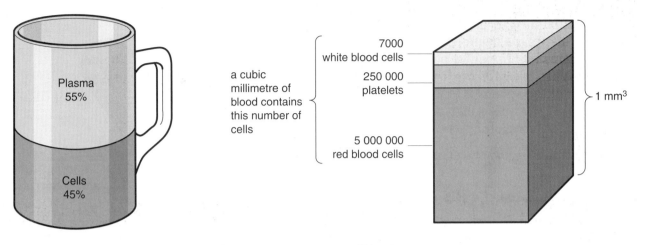

Figure 1.3.2 The percentage composition of the various components of blood

diseases. Plasma also contains an enzyme called **prothrombin** which is produced in the liver in the presence of **vitamin K**. This enzyme functions during blood clotting.

The proteins in blood are important to give your blood its thickness or **viscosity**. This helps to keep up the proper pressure in the blood vessels. The other materials in plasma are:

- **Inorganic materials**, dissolved in water. These compounds include carbonates, chlorides, and phosphates of the elements calcium, sodium, magnesium and potassium.

They are necessary for normal functioning of your body tissues. Without calcium compounds, for example, your blood would not clot in a wound.

- **Digested foods** in the form of glucose, fatty acids, glycerol, and amino acids. These are carried to body tissues, the liver, and other storage areas.
- **Nitrogenous wastes** from protein metabolism in the tissues. One of these wastes is **urea** which is made in the liver during the breakdown of unwanted amino acids. These nitrogenous wastes travel in the plasma to the **organs of excretion** (see page 62).

The blood cells

There are three solid components of blood. They are **red blood cells (erythrocytes)**, **white blood cells (leucocytes)**, and **platelets (thrombocytes)**, which are not complete cells like the red and white cells. The red blood cells are biconcave discs. Ten million of them can be spread out in about 6 square centimetres. A normal person has about 5.5 million per cubic millimetre of blood. If you laid the cells side by side, that is enough to go around the world four times (Figure 1.3.3).

The red pigment in red cells is a protein called **haemoglobin**. The red cells are produced in the red **bone marrow** at the ends of such bones as the ribs, vertebrae and skull. In children, even the ends of the long limb bones can make these cells. Developing red cells are large, colourless,

Figure 1.3.3 Your red blood cells would encircle the world four times

and have big nuclei. Usually they have lost the nuclei by the time they enter the bloodstream. They have also made haemoglobin by this time which remains functional for between 20 and 120 days. When red cells die, they are removed by the liver or the **spleen**. At this time, these organs recycle valuable substances such as iron from the old haemoglobin to be used to make new haemoglobin.

What do red blood cells do?

The pigment, haemoglobin, is a complex protein containing iron and which fills the red blood cells. Haemoglobin can carry oxygen because it contains iron. You have probably seen iron turn red with rust when iron is **oxidised**. This means that iron has combined with oxygen in the air. The iron in haemoglobin combines with oxygen in your lungs. Rusty iron does not give up its oxygen easily but the iron of haemoglobin does. In fact, haemoglobin releases oxygen at just the right time and place in the body. The red cells are bright red when haemoglobin combines with oxygen (**oxyhaemoglobin**). The oxygen is carried to tissues that need it and is given up. The red cells also play a part in **carbon dioxide transport**.

The white blood cells

Most white cells are larger than red cells. They differ from them in three ways:

- White blood cells have nuclei.
- White blood cells have no haemoglobin. They are almost colourless.

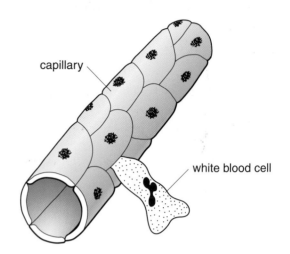

Figure 1.3.4 White blood cells can just squeeze through capillaries

- Some white blood cells are capable of movement.

The ratio of red blood cells to white blood cells is about 600:1. White blood cells are formed in red bone marrow and in the **lymph glands**. Normally, you have about 8 000 white blood cells in a cubic millimetre of blood.

Some white blood cells move about on their own. They can squeeze through capillary walls into the tissue spaces (Figure 1.3.4). Here, they engulf solid particles and bacteria. Thus, white blood cells are important in defending your body against infection. Whenever you develop an infection, your white cell count increases. It may rise from 8 000 to more than 25 000 per cubic millimetre. These cells collect in an infected area. Then they ingest and destroy bacteria.

The platelets

Platelets are much smaller than red blood cells. Platelets are irregularly shaped and colourless. They are formed in the red bone marrow and are important in **blood clotting**.

Clotting results from chemical and physical changes in the blood. When a blood vessel is cut, platelets break down in the blood which leaves the vessel. Platelets are probably destroyed by tissue fluids at the wound site. In the process, they release **thromboplastin**. This reacts with **prothrombin** in the presence of calcium to form **thrombin**. Thrombin changes fibrinogen, a blood protein, to **fibrin**. Fibrin is made of tiny threads which form a network that traps blood cells. When this happens, a clot is formed that stops any more blood from escaping. The trapped blood cells dry out to form a scab.

Summary of clotting:

> 1 thromboplastin + calcium + prothrombin
> $\rightarrow$ thrombin
> 2 thrombin + fibrinogen $\rightarrow$ fibrin

If any of these substances is missing, blood will not clot.

Table 1 Summary of the composition of the blood

Plasma	Solids
water	red cells
fibrinogen	white cells
serum albumin	platelets
globulin	
digested foods (glucose, fatty acids, glycerol, amino acids)	
mineral salts	
vitamins	
cell wastes (urea, carbon dioxide)	
hormones	

Table 2 Blood as a transporting medium

Transportation of	From	To	For the purpose of
digested food	alimentary canal and liver	tissues	growth and energy release
cell wastes	tissues	lungs, skin, and kidneys	excretion
water	alimentary canal	kidneys and skin	excretion and balancing concentration of body fluids
oxygen	lungs	tissues	respiration
heat	tissues	skin	regulation of body temperature
hormones	ductless glands	tissues	regulation of body activities

Blood types

Blood type is determined by the presence or absence of certain proteins in the red blood cells and plasma. Such proteins in the red cells are called antigens. If the antigen A is present, then an individual is said to have type A blood. Type B blood contains antigen B; AB-type blood contains both A and B antigens. Type O blood has neither of the antigens present. Blood types are the result of genetic effects (see page 151.)

Substances called **antibodies** may also be present in the plasma. These can react with the antigens causing the red blood cells to clump together. The antibodies found in a particular blood type are always the opposite of the antigen. For example, group A blood contains the A antigen but the b antibody. Group B blood has the B antigen and the a antibody. AB blood contains both antigens but neither antibody, while group O blood has no antigens but contains both a and b antibodies.

LIFE PROCESSES

35

Circulation

If antigens and antibodies of the same type come together, they cause the red cells to clump (**agglutination**). Thus, if two samples of blood of the same type are mixed together, no clumping occurs, whereas if different blood types are mixed, the cells clump. If this should happen as a result of incorrect matching of blood during transfusion, the result could be fatal.

Table 3 Human blood types

Type of blood	Antigens in red cells	Antibodies in plasma
A	A	b
B	B	a
AB	A and B	neither a nor b
O	neither A nor B	a and b

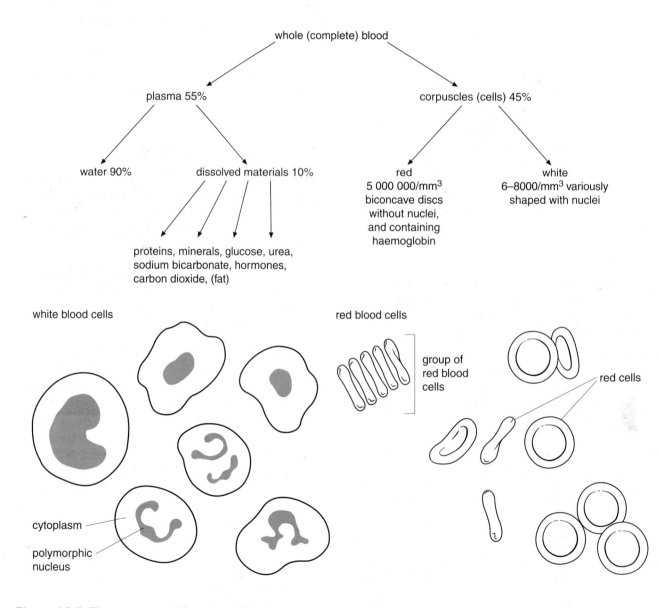

Figure 1.3.5 The structure and functions of blood

Blood typing

A sample of blood is mixed with a drop of **serum** (plasma without clotting factors) containing a antibodies. Another drop of the sample blood is mixed with a drop of serum containing b anitbodies. Whether or not the red blood cells of the sample clump together, will quickly serve to identify the type of blood contained in the sample, because clumping would indicate the presence of A or B antigens.

Table 5 The percentage of blood types in UK populations

Population	O	A	B	AB
English	46	42	8	4
Welsh	48	37	10	5
Scots	50	32	13	5
Irish	52	30	13	5

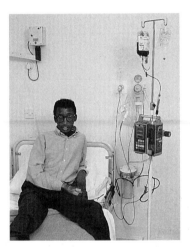

This man has sickle cell anaemia and is having a blood transfusion which could be life-saving

People with group AB are called **universal recipients** and can, theoretically, receive blood from any other group because this blood type lacks antibodies. However, it is not quite as simple as this because there are other factors which also determine the success of transfusions. Blood from a donor is therefore always cross-matched with a recipient's blood to be sure they are compatible.

The Rh factor in blood

The A, B, AB and O method of blood grouping is just one of several. The **Rh factor** is another. Rh is the shortened form of **Rhesus factor**, named after the rhesus monkey in which it was discovered (see photo overleaf).

About 85% of humans have the Rh factor in their blood and are called Rh positive. The remainder are Rh negative. Like blood groups, the Rh factor is inherited.

Before a transfusion takes place it must be established that the antigens in the donor's blood will not cause clumping of the recipient's blood, which might block important blood vessels (Figure 1.3.6). In emergencies, group O blood can be transfused into most people without adverse effects because it contains no antigens. People with blood type O are thus called **universal donors**.

If Rh negative patients receive Rh positive blood, they produce antibodies against this factor because it is an antigen. These antibodies cause the red blood cells of the Rh positive blood to agglutinate and dissolve. There is little

Table 4 Reactions when blood is mixed with serums containing a and b antibodies

Sample mixed with a antibodies	Sample mixed with b antibodies	Antigens present	Blood type
no clumping	no clumping	none	O
no clumping	clumping	B	B
clumping	no clumping	A	A
clumping	clumping	A and B	AB

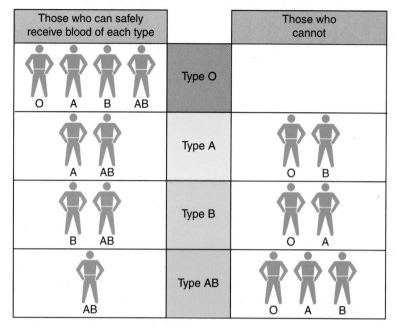

Those who can safely receive blood of each type		Those who cannot
O A B AB	Type O	
A AB	Type A	O B
B AB	Type B	O A
AB	Type AB	O A B

Blood group of recipient	Donor's blood group			
	Group O	Group A	Group B	Group AB
Group O	✓	●	●	●
Group A	✓	✓	●	●
Group B	✓	●	✓	●
Group AB	✓	✓	✓	✓

✓ Safe transfusion (compatible bloods)

● Dangerous transfusion (incompatible bloods)

Figure 1.3.6 Blood transfusions

danger during the first transfusion. This is because the antibody is not present when the Rh positive blood is added. However, a second transfusion can be serious or even fatal because the patient has already formed antibodies against the Rh positive blood.

The Rh factor and childbirth

The Rh factor may cause a problem during pregnancy in about one in every 300–400 mothers (Figure 1.3.7). It may happen when the mother is Rh negative and the father is Rh positive. The child may inherit the Rh positive factor from the father. Rh positive blood may seep from the child into the mother's circulation. It can go through the membranes that normally keep the two circulations separate. The mother's blood can also seep into the child in the same way. The seepage does not happen very often, allowing many Rh negative mothers to have Rh positive children. Usually the situation only becomes serious if seepage occurs again with a second Rh positive child. In

Rhesus monkeys

An investigation into compatibility of simulated blood groups

For this investigation you will need to make up solutions which simulate various blood groups and which simulate agglutination. Materials you will need:

A white tile, a glass tube, eight stock solutions representing blood groups.

These are:

Donor groups

O – distilled water and eosin
A – $FeCl_3$ and eosin
B – K_2SO_4 and eosin
AB – $FeCl_3$ plus K_2SO_4 plus eosin

Recipient groups

O – $BaCl_2$ plus NaOH plus eosin
A – $BaCl_2$ plus eosin
B – NaOH and eosin
AB – distilled water and eosin

Procedure

Using the glass tube, spot samples of the simulated blood on to the tile according to the following instructions. Wash the tube carefully between each sample withdrawn, or the stock solutions will become contaminated and useless. Mix the following solutions on the tile:

1 Donor A + Recipient A
2 Donor A + Recipient B
3 Donor A + Recipient AB
4 Donor A + Recipient O

In each case explain the observation.

Wash the tile thoroughly and begin the investigation again using Donor B instead of Donor A. Explain your observations.

Wash everything thoroughly again. Repeat using Donor AB and Donor O. From your observations deduce:

a) Which donor group can mix with any other group?

b) Which recipient group can mix with any other?

c) Why is it important that the tile should be washed thoroughly?

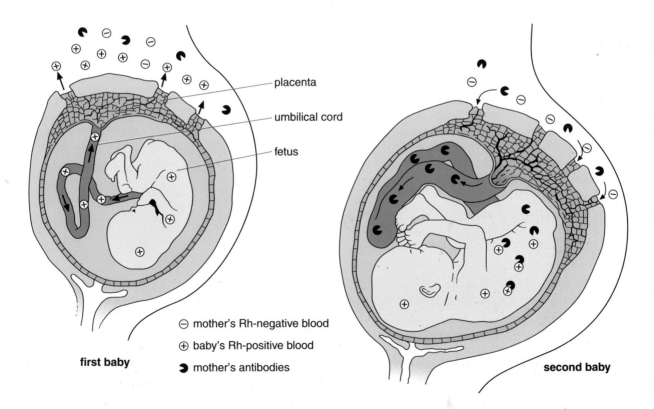

○ mother's Rh-negative blood
⊕ baby's Rh-positive blood
❱ mother's antibodies

first baby

second baby

Figure 1.3.7 If a Rh- mother produce Rh+ antibodies as a result of an earlier Rh+ fetus, a problem may arise if she becomes pregnant with another Rh+ fetus

that case, the mother has already produced antibodies from her first pregnancy.

When the mother's blood seeps into the second child's blood, these antibodies can cause serious damage. Sometimes the child dies before birth, but often the damage is not so serious. A transfusion immediately after birth may save the child's life. The child's blood may be almost entirely replaced by tranfused Rh negative blood.

The usual method to avoid the Rh problem in future pregnancies is to give the Rh negative mother an injection a few days after the time of birth of her first baby. This injection contains Rh antibodies. These antibodies circulate in the mother's blood for several weeks. They destroy all the Rh positive factor from the baby. Thus, the mother's system does not develop its own antibodies against Rh factor. This is a form of passive immunity and the mother should not have Rh problems with future babies.

The circulatory system

The heart

The heart is a pear-shaped muscular organ, enclosed in a sac called the **pericardium**. It is located under your breastbone and between your lungs. Your heart is made of two sides, right and left. These two halves are separated by a wall called the **septum**. Each half has two chambers. There is a thin-walled chamber called the **atrium** and there is a thick, muscular **ventricle**. The two atria are reservoirs for the blood that enters the heart. They contract at the same time so that the pressure in them increases. This forces the blood into the two ventricles. The muscular walls of the ventricles contract and the resulting increase in pressure forces the blood out through the main arteries.

The heart has two sets of **one-way valves**. They control the flow of blood from the ventricles. The valves between the atria and venrtricles are called **atrioventricular valves (the a-v valves)**. These are like flaps and are anchored to the floor of the ventricles by tendon-like strands. Blood passes freely through the a-v valves into the ventricles, but these valves cannot be opened from the lower side because the tendons anchor them. This results in blood not being able to flow backward into the atria when the ventricles contract.

The other valves are called the **semilunar valves (s-l valves)**. These are cup-shaped and are located at the openings of the arteries. The force of blood passing from the ventricles into the arteries opens the s-l valves which prevent blood from moving back into the ventricles.

The passage of blood through the heart

One of the best ways to learn about the parts of the heart is to trace the path of blood as it passes through it (Figure 1.3.8). Blood, which has given up most of its oxygen, first enters the right atrium from two different directions. Blood enters through the **superior vena cava**, from the head and upper parts of the body. It enters through the **inferior vena cava** from the lower parts of the body.

From the right atrium, the blood goes through the right a-v valve into the right ventricle. Then the right ventricle contracts and forces blood through a set of s-l valves into the **pulmonary artery** which takes the blood to the lungs. After picking up oxygen in the lungs, **pulmonary veins** carry the oxygenated blood to the left atrium. From there, the blood passes through the left a-v valve into the left ventricle. Finally, the blood passes out through the aorta and goes to all parts of the body apart from the lungs, under pressure.

The heart's muscle cells are supplied by special arteries called **coronary arteries** (Figure 1.3.9). The right and left coronary arteries curve downward around each side of the heart. Each branches to smaller vessels that penetrate the heart's muscle tissue.

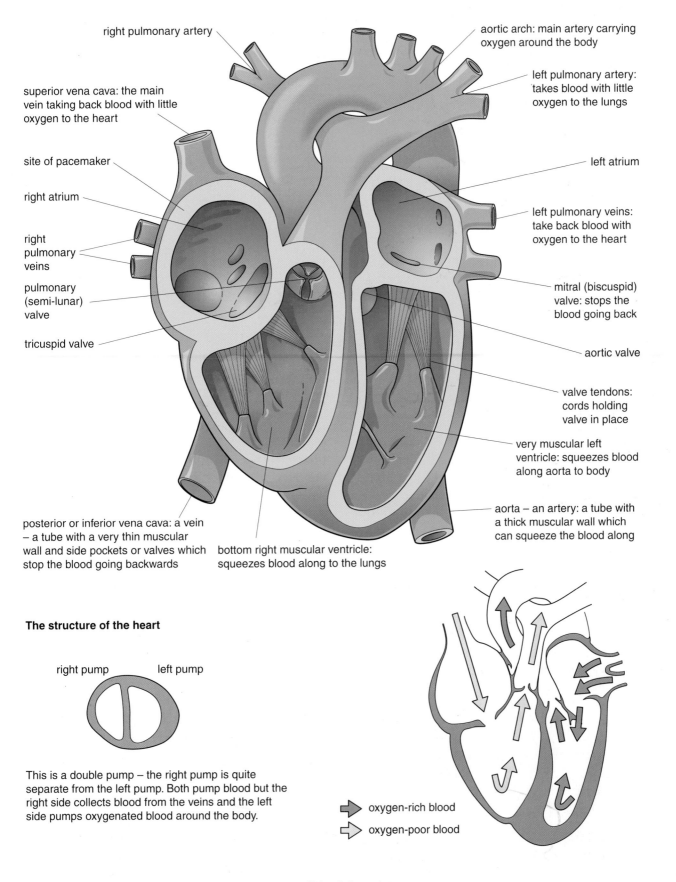

Figure 1.3.8 Section through the heart and the passage of blood through it

The labels on the figure read:

- right pulmonary artery
- aortic arch: main artery carrying oxygen around the body
- superior vena cava: the main vein taking back blood with little oxygen to the heart
- left pulmonary artery: takes blood with little oxygen to the lungs
- site of pacemaker
- left atrium
- right atrium
- left pulmonary veins: take back blood with oxygen to the heart
- right pulmonary veins
- pulmonary (semi-lunar) valve
- mitral (biscuspid) valve: stops the blood going back
- tricuspid valve
- aortic valve
- valve tendons: cords holding valve in place
- very muscular left ventricle: squeezes blood along aorta to body
- posterior or inferior vena cava: a vein – a tube with a very thin muscular wall and side pockets or valves which stop the blood going backwards
- bottom right muscular ventricle: squeezes blood along to the lungs
- aorta – an artery: a tube with a thick muscular wall which can squeeze the blood along

The structure of the heart

right pump   left pump

This is a double pump – the right pump is quite separate from the left pump. Both pump blood but the right side collects blood from the veins and the left side pumps oxygenated blood around the body.

oxygen-rich blood
oxygen-poor blood

LIFE PROCESSES

LIFE PROCESSES

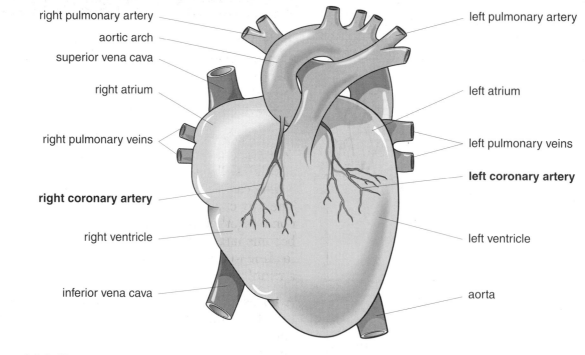

right pulmonary artery

aortic arch

superior vena cava

right atrium

right pulmonary veins

right coronary artery

right ventricle

inferior vena cava

left pulmonary artery

left atrium

left pulmonary veins

left coronary artery

left ventricle

aorta

Figure 1.3.9 The coronary arteries

The heart as a highly effective pump

A complete cycle of heart activity is called a **beat** and has two **phases**. In the first phase, or **systole**, the ventricles contract and force blood into the arteries. In the second phase, or **diastole**, the ventricles relax and take in blood from the atria.

A normal heart beat makes 'lub' and 'dup' sounds, repeated over and over. They are in perfect rhythm, the 'lub' being the systole phase. It is the sound of the contraction of the ventricles muscles and the closing of the a-v valves. The 'dup' is the diastole and is the sound of the closing of the semi lunar valves.

The heart of an average adult beats about 70 times per minute. This is when the person is resting. During physical exercise, the heart rate may be as high as 180 beats per minute.

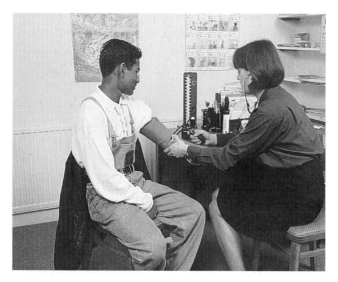

This doctor is measuring a patient's blood pressure

The doctor presses the patient's radial artery in her wrist to take her pulse

An investigation of the effect of exercise on pulse rate

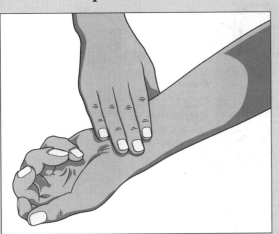

Figure 1.3.10 How to take a pulse

Procedure

Using your finger tip, find your pulse where an artery passes over a bone near to the surface of the skin in your wrist (see Figure 1.3.10). Do not use your thumb because you have a pulse in it which might confuse your results. Count the number of times your pulse beats in 15 seconds and record the results.

Repeat three times and find the average. Multiply this figure by four to find your pulse rate per minute.

IF THERE IS ANY MEDICAL REASON WHY YOU SHOULD NOT EXERCISE, DO NOT PROCEED ANY FURTHER.

Exercise for five minutes by press-ups or by stepping rapidly on and off a box. Really put yourself to the test but not for more than five minutes. Sit down and find your pulse again.

Record your pulse every 15 seconds until it returns to normal. Plot your pulse rate per minute against time after exercise in seconds. From your observations deduce:

a) Your resting pulse rate per minute

b) The length of time taken for your pulse to return to normal.

Suggest why your pulse rate did not return to normal as soon as you stopped exercising.

The blood vessels

Blood moves in a series of tubes of different sizes. **Arteries** and **arterioles** carry blood away from the heart. **Veins** and **venules** carry blood towards the heart. **Capillaries** are very small, thin-walled vessels (Figure 1.3.11).

The aorta branches into several smaller arteries. These further branch and become arterioles. The arterioles branch into capillaries which can only be seen using a microscope. The tiny capillaries pass through tissues and come together to form venules. These join to become larger and larger veins, eventually leading to the venae cavae which enter the right atrium.

Arteries have muscular and elastic walls with smooth linings. This allows arteries to expand and absorb great pressure created by contracting ventricles at systole. The pulse you feel in your wrist is caused by systolic pressure in an artery and has the same rhythm as your heartbeat. The elastic walls also maintain the pressure in the arteries when the ventricles relax. This is called **diastolic pressure**. The pressure exerted by the left ventricle is much higher than that of the right ventricle because it has to force the blood further.

What are the capillaries?

Arterioles penetrate the tissues. Here they branch into capillaries. The walls of capillaries are only one cell thick. Also, capillaries are only a little larger in diameter than red blood cells. So red blood cells must pass through the capillaries in single file. Sometimes the red cells are even pressed out of shape by the capillary walls. Veins and arteries are important for carrying blood through our bodies but the vital exchanges between blood and tissues occur because of our capillaries.

Dissolved products of digestion, wastes, and gases pass freely through their thin walls. White cells squeeze through tiny gaps in the walls. This is how white cells leave the bloodstream and enter the tissue spaces. Also, some of the plasma diffuses from the blood through the capillary walls. This is **tissue fluid** (Figure 1.3.12).

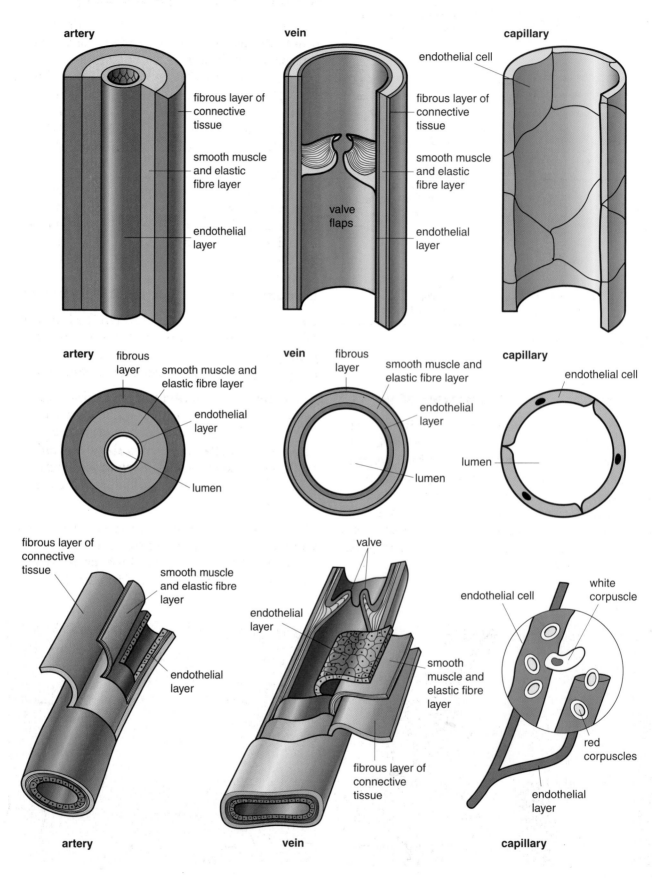

artery

fibrous layer of connective tissue

smooth muscle and elastic fibre layer

endothelial layer

vein

endothelial cell

fibrous layer of connective tissue

smooth muscle and elastic fibre layer

valve flaps

endothelial layer

capillary

artery fibrous layer

smooth muscle and elastic fibre layer

endothelial layer

lumen

vein fibrous layer

smooth muscle and elastic fibre layer

endothelial layer

lumen

capillary

endothelial cell

lumen

fibrous layer of connective tissue

smooth muscle and elastic fibre layer

endothelial layer

artery

valve

endothelial layer

smooth muscle and elastic fibre layer

fibrous layer of connective tissue

vein

endothelial cell

white corpuscle

red corpuscles

endothelial layer

capillary

Figure 1.3.11 The structure of blood vessels

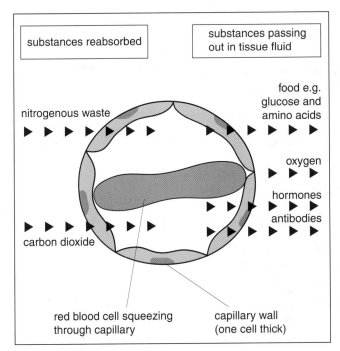

substances reabsorbed

substances passing out in tissue fluid

nitrogenous waste

food e.g. glucose and amino acids

oxygen

hormones

antibodies

carbon dioxide

red blood cell squeezing through capillary

capillary wall (one cell thick)

Figure 1.3.12 The formation of tissue fluid

The structure and function of veins

As capillaries leave a tissue, they unite to form venules, then veins. Most veins carry dark red blood which has given up some of its oxygen to the tissues. You can see some veins through your skin as blue lines. The walls of veins are much thinner and less muscular than those of arteries. The diameter of a vein is greater than a corresponding artery. Many veins have cup-like valves which keep blood from flowing backwards.

Circulation through the body

Our four-chambered heart is really a **double pump**. Its two sides work together. Each side pumps blood through a major division of our circulatory system. The right side pumps blood through the **pulmonary circulation**. This is the circulation of blood through the lungs. The left side pumps blood in the **systemic circulation** through the rest of the body (see overleaf).

The right side of your heart receives deoxygenated blood from the body. It pumps this dark red blood through arteries to the pulmonary circulation. The **pulmonary artery** branches to supply both lungs. Here, the blood gives up its carbon dioxide and receives oxygen. The oxygenated blood is bright red and is returned to the heart in the **pulmonary veins**.

Oxygenated blood passes through the left side of the heart and out through the aorta under great pressure to the systemic circulation, supplying all the organ systems of your body.

The **coronary circulation** supplies the heart muscle. A blood clot in a coronary artery may stop flow of blood to the heart muscle and cause a 'heart attack'. The **renal circulation** supplies the kidneys with blood containing urea to be filtered out in order to form urine (see page 66).

The **portal circulation** is an extensive system of veins which lead from the small intestine and stomach. They join to form the **hepatic portal vein** which takes digested food to the liver.

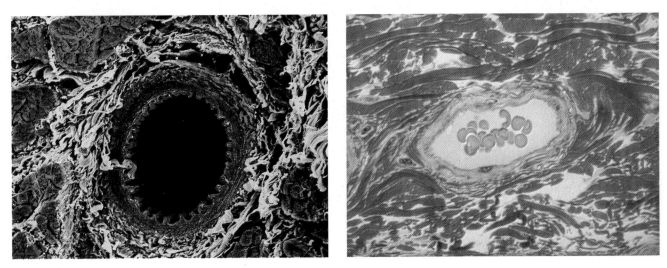

An artery (left) is much thicker and narrower than a vein (right)

LIFE PROCESSES

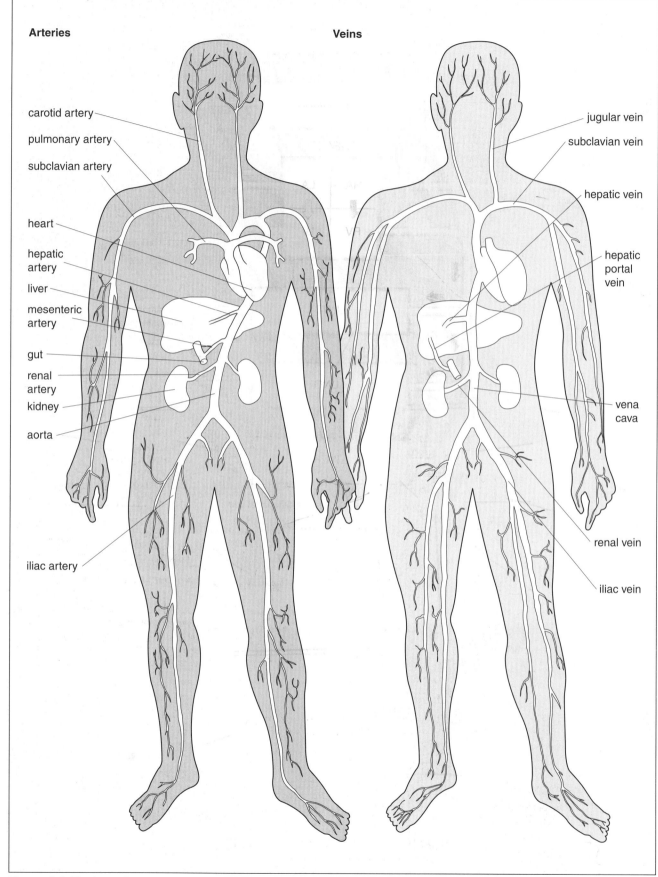

Arteries

Veins

carotid artery

pulmonary artery

subclavian artery

heart

hepatic
artery

liver

mesenteric
artery

gut

renal
artery

kidney

aorta

iliac artery

jugular vein

subclavian vein

hepatic vein

hepatic
portal
vein

vena
cava

renal vein

iliac vein

Figure 1.3.13 Arterial and venous circulations

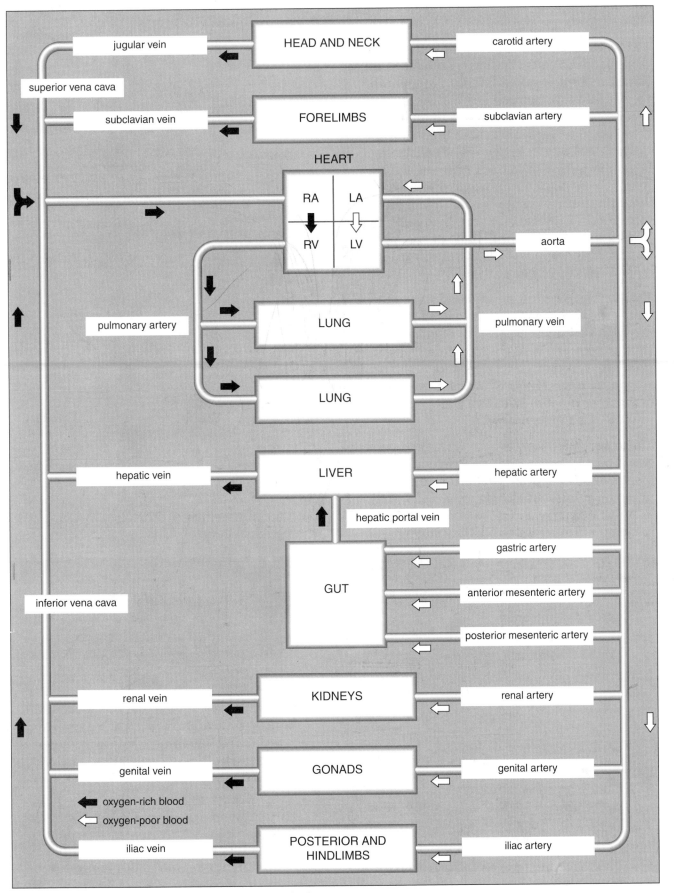

Figure I.3.14 A plan of blood circulation through the body

The lymph

The **tissue fluid** that bathes your cells by seeping out of your capillaries collects in tubes. Here, the fluid is called **lymph**. Tiny lymph vessels join together and form larger ones. **Lymph nodes** are enlargements in the lymph vessels where the lymph tubes branch into many fine vessels. Certain white cells collect here and destroy bacteria if they are in the lymph. Your neck, armpits and groin have the largest number of lymph nodes. If you have an infection in your hand or arm, the lymph nodes in your armpit often swell and become painful. Your tonsils and adenoids in your throat are collections of lymphatic tissue.

The lymph in your right arm and right side of your head and neck enters a large vessel called the **right lymphatic duct**. This opens into the **right subclavian vein** which comes from your right arm. The lymph thus returns to the blood stream. The lymph from the rest of your body drains into the **thoracic duct** which empties into the **left subclavian vein**.

The larger lymph vessels have valves in their walls which control the flow of lymph in one direction. Many lymph vessels are wrapped around the pulsating arteries running between and within muscles. When these muscles contract, they force the lymph to move under pressure and the

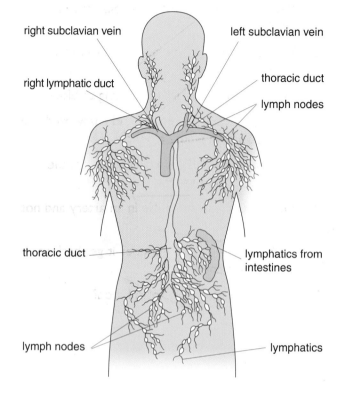

direction of movement is controlled by the valves which act like those of veins.

Figure 1.3.15 The lymphatic system

Summary

1 Your circulatory system transports blood through your body.

2 Blood is a fluid tissue. It is made up of liquid plasma and solid components.

3 Plasma contains water, blood proteins, prothrombin, inorganic substances, digested foods, and cell wastes.

4 Blood has three solid parts: red cells, white cells and platelets.

5 Red cells transport oxygen to the body cells. They also transport carbon dioxide away from the cells as a waste product.

6 White cells help fight disease-causing bacteria.

7 Platelets are important in the process of blood clotting.

8 The heart is a pump that sends blood to all parts of your circulatory system. It has two atria and two ventricles. The atria receive blood from veins. The ventricles force blood through the arteries by contracting.

9 Arteries carry blood from the heart to the tissues. Veins return it to the heart. The arterial and venous systems are connected by networks of tiny capillaries. The real interaction of blood and cells takes place in these capillaries.

10 Part of the blood plasma seeps into tissue spaces. From there, it is collected into special tubes as lymph. The fluid is filtered by the lymph nodes and returned to the bloodstream.

Questions for review

1 What is blood?

2 Where are the various blood cells made in the body?

3 What conditions might cause the white blood cell count to go up?

4 What are the stages in the clotting of blood?

5 Why is plasma more quickly and easily used in a transfusion than whole blood?

6 Trace the path of a drop of blood from the right atrium to the aorta.

7 Why can you feel the pulse in an artery and not in a vein?

8 What is tissue fluid? How does it get back to the bloodstream?

9 How does lymph differ from blood?

Applying principles and concepts

1 What is the reason for the saying that 'we are as young as our arteries'?

2 Alcohol dilates the arteries in the skin. What would be its effect on the temperature control of the body?

3 In a Rh negative patient, why might a second transfusion with Rh positive blood be fatal, eventhough the first transfusion with Rh positive blood caused no complications?

4 How is recycling of resources by the body demonstrated in the manufacture of red blood cells?

LIFE PROCESSES

Breathing

Learning Objectives

By the end of this chapter, you should be able to:

- Understand the meaning of the term, breathing
- Describe the functions of the parts of the breathing system

- Explain gas exchange in the lungs
- Define oxygen debt
- Understand the importance of healthy lungs

The two phases of breathing

Breathing is the mechanical process that allows air in and out of the lungs so that oxygen can diffuse into the blood and so that carbon dioxide and water can diffuse out.

The organs involved in breathing can be divided into two groups. The first group includes the passages through which air travels to get into the bloodstream. These are the nostrils, nasal passages, throat, trachea, bronchi, bronchioles and air sacs. The second includes those organs involved in the mechanics of breathing. This group includes the ribs, rib muscles, diaphragm, and abdominal muscles.

The breathing system

The nose and nasal passages

The air enters the nose in two streams through two **nostrils**. From the nostrils, air enters the **nasal passages** lying above the mouth cavity. Before air enters the nasal passages, however, hairs and most mucous membranes screen out dirt. A wafer-thin network of bones acts to warm up the air by friction as the molecules of gases pass through the nasal passages. These advantages are lost when you breathe in through your mouth.

The trachea

From the nasal passages air goes through the **throat (pharynx)** and down the **windpipe** (trachea) (Figure 1.4.2). The upper end of the trachea is protected by a flap of cartilage called the **epiglottis**. When you swallow, the epiglottis closes over the trachea. This prevents food from getting into the lungs. The upper end of the trachea holds the **voice box (larynx)**. Vocal cords are located inside the larynx and are used to make sounds when they vibrate. By learning to control the vibrations we learn to talk.

Horseshoe-shaped bands of cartilage support the walls of the trachea to keep it open for the passage of air. The trachea and its branches are lined with tiny, constantly moving, hair-like **cilia**. They carry dirt and other foreign particles that we inhale upward toward the mouth. This

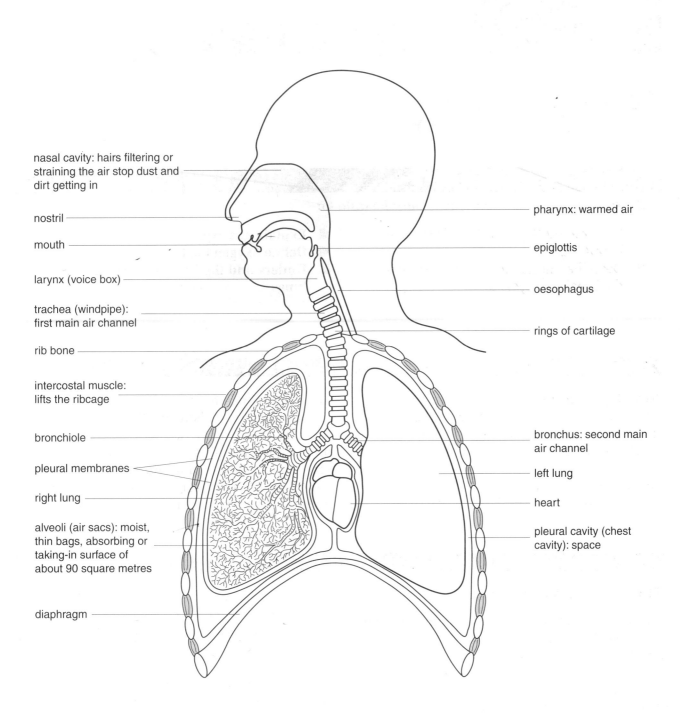

nasal cavity: hairs filtering or straining the air stop dust and dirt getting in

nostril

mouth

larynx (voice box)

trachea (windpipe): first main air channel

rib bone

intercostal muscle: lifts the ribcage

bronchiole

pleural membranes

right lung

alveoli (air sacs): moist, thin bags, absorbing or taking-in surface of about 90 square metres

diaphragm

pharynx: warmed air

epiglottis

oesophagus

rings of cartilage

bronchus: second main air channel

left lung

heart

pleural cavity (chest cavity): space

Figure 1.4.1 The breathing system

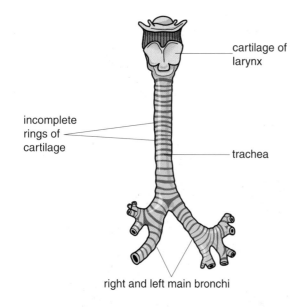

Figure 1.4.2 The trachea

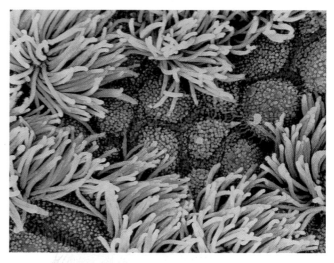

Ciliated epithelium

dirt is removed when you cough, sneeze or swallow.

The bronchi and air sacs

The trachea divides at its lower end into two branches called **bronchi**. One bronchus extends to each lung and then divides to many smaller **bronchioles**, ending in clusters of tiny air sacs called **alveoli** (Figure 1.4.3). The walls of the alveoli are very thin and elastic, enabling exchange of gases. The oxygen in air cannot travel any further than the alveoli without the help of the blood system. Capillaries carry the blood to the alveoli so that oxygen can be collected and carbon dioxide unloaded. All the tiny alveoli provide an enormous surface area in contact with air for **gaseous exchange**. Thus the lungs provide oxygen for the blood to carry to the millions of cells that need it for respiration (see page 8).

The lungs fill the chest cavity from under the shoulders down to the diaphragm – except for

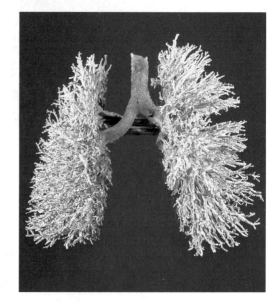

The respiratory tree

the space occupied by the heart, trachea, oesophagus and blood vessels. Lungs are made mainly of bronchioles, air sacs, and blood vessels held together by **connective tissue**. This structure makes the lungs spongy. Surrounding the lungs are the **pleural membranes** which secrete mucus for lubrication. This allows them to move freely in the chest during breathing.

Breathing movements

When you inhale, your chest bulges and you might think that this is because your lungs draw in air and expand. Actually, this is the opposite of what happens (Figure 1.4.4). The lungs, although elastic in nature, have no muscle so cannot expand or contract on their own. They are spongy, air-filled sacs in the chest cavity. The power for breathing comes from rib,

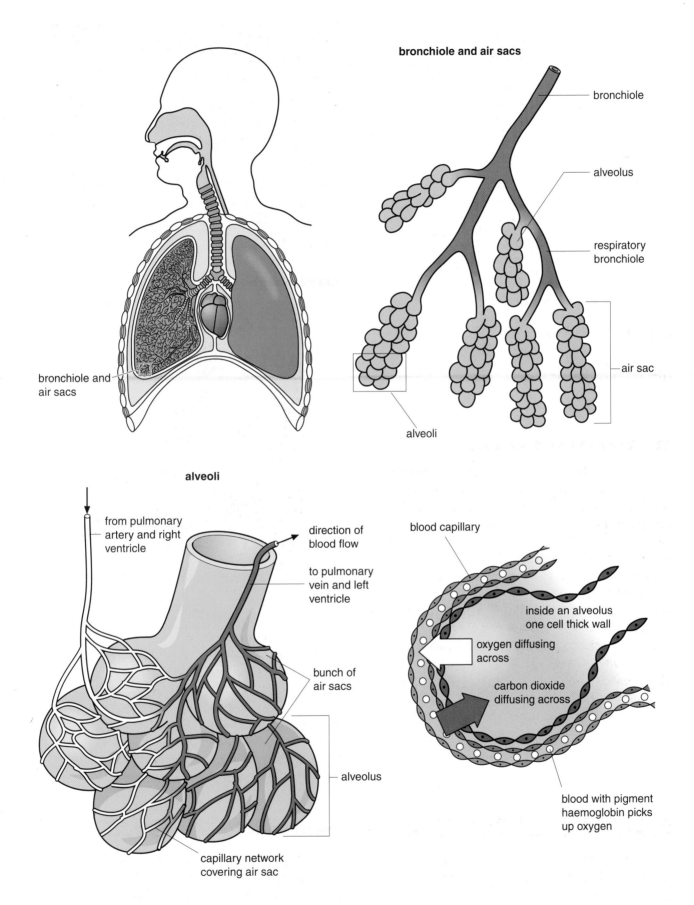

bronchiole and air sacs

bronchiole

alveolus

respiratory bronchiole

air sac

alveoli

bronchiole and air sacs

alveoli

from pulmonary artery and right ventricle

direction of blood flow

to pulmonary vein and left ventricle

bunch of air sacs

alveolus

capillary network covering air sac

blood capillary

inside an alveolus one cell thick wall

oxygen diffusing across

carbon dioxide diffusing across

blood with pigment haemoglobin picks up oxygen

Figure 1.4.3 The structure of bronchioles and air sacs

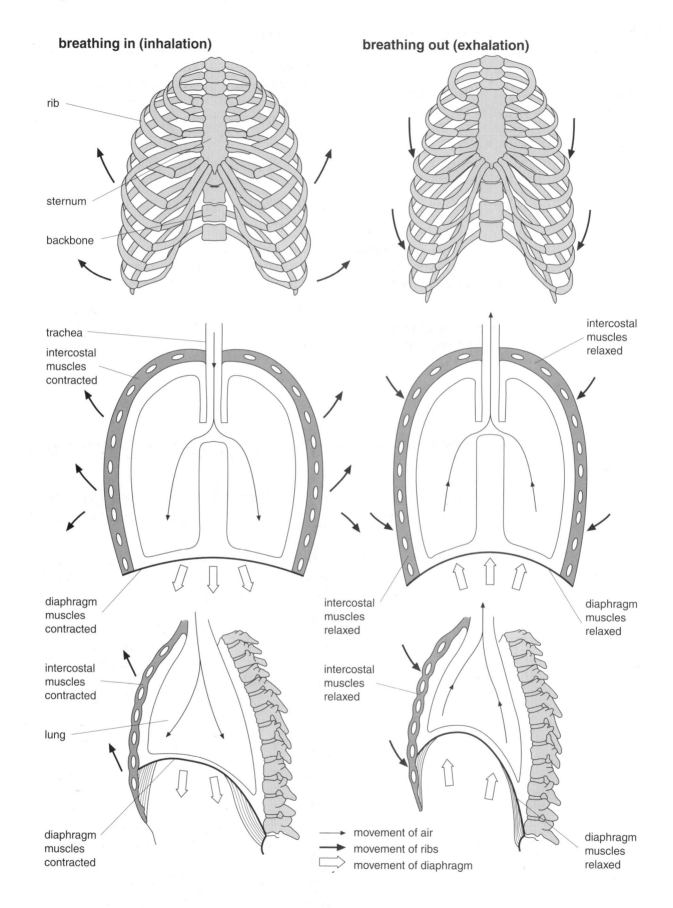

breathing in (inhalation)

breathing out (exhalation)

rib

sternum

backbone

trachea

intercostal muscles relaxed

intercostal muscles contracted

diaphragm muscles contracted

intercostal muscles relaxed

diaphragm muscles relaxed

intercostal muscles contracted

intercostal muscles relaxed

lung

diaphragm muscles contracted

diaphragm muscles relaxed

→ movement of air

➜ movement of ribs

⇨ movement of diaphragm

Figure 1.4.4 Breathing action

diaphragm, and abdominal muscles. These muscles control the size and air pressure in the chest cavity.

The intake of air, or **inspiration**, occurs when the chest cavity expands. When this happens, air pressure inside the sealed cavity decreases. Air rushes into the lungs to equalise this pressure. The following movements are involved in expanding the chest cavity:

1 The rib muscles contract, pulling the ribs up and out.

2 The muscles of the dome-shaped diaphragm contract. This straightens and lowers the diaphragm, enlarging the chest cavity from below.

3 The abdominal muscles relax, allowing compression of the abdomen when the diaphragm lowers.

Expiration, or forcing air from the lungs, occurs when the chest cavity is reduced in size. This increases the air pressure inside the cavity. In order to equalise internal and external air pressure, air is forced out of the lungs. The following movements are involved in expiration when you exhale:

1 The rib muscles relax allowing the ribs to spring back.

2 The diaphragm relaxes, rising to its original position.

3 The abdominal muscles contract, pushing the abdominal organs up against the diaphragm.

4 The elastic lung tissues were stretched when the lungs were full. These recoil and force air out of the lungs.

The control of breathing

Both nerves and chemicals control the depth and rate of breathing. Nerves from the lungs, diaphragm, and rib muscles lead to a control centre at the base of the brain (**the brain stem**). It controls the regular rhythm of breathing as it detects the amount of carbon dioxide in the blood. If the carbon dioxide concentration is high, the brain signals the diaphragm and rib muscles to increase the breathing rate. This increased rate forces more carbon dioxide out through the lungs and lowers the concentration in the blood. Breathing settles back to a normal rate. In humans, inspiration and expiration occur from 16 to 24 times a minute. The exact rate depends on physical activity, position, mood and age.

The air capacity of the lungs

Only about 500 cm³ of air are involved each time we inhale and exhale (Figure 1.4.5). The air involved in normal, relaxed breathing is called **tidal air**. Forced breathing increases the amount of air movement.

You can demonstrate the effects of forced

SUGGESTED INVESTIGATIONS

An investigation of the effect of exercise on your rate of breathing

Procedure

While resting, place your hands on the lower part of your rib cage and count the number of inspirations over a period of thirty seconds. For this to be valid, you must breathe naturally – do not force your breathing. Repeat this four times so that you can calculate your average rate of breathing at rest. Record the average reading and double it to find your breathing rate per minute.

Run on the spot vigorously for one minute, stop and repeat the first procedure. Again repeat four times to find an average reading for your rate of breathing after exercise. Draw a bar chart to show the change in breathing rates of the students in a class after exercise.

What is your average breathing rate before and after exercise? How close is your average to the average for the class?

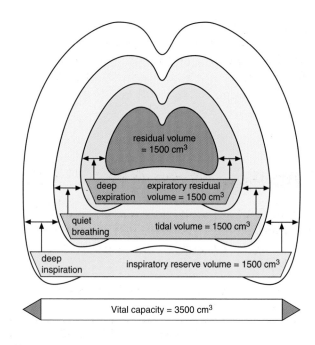

residual volume
= 1500 cm³

deep
expiration

expiratory residual
volume = 1500 cm³

quiet
breathing

tidal volume = 1500 cm³

deep
inspiration

inspiratory reserve volume = 1500 cm³

Vital capacity = 3500 cm³

Figure 1.4.5 The air capacity of the lungs

breathing. Inhale normally without forcing. Your lungs now contain about 2 800 cm³ of air. Now exhale normally. You have moved 500 cm³ of tidal air from the lungs. Now, without inhaling again, force out all the air you can. You have now exhaled an additional 1 100 cm³ of **supplemental air**. The lungs now contain about 1 200 cm³ of **residual air**, which you cannot force out.

When you inhale normally again, you replace the supplemental and tidal air. If you inhale with force, you can add about 3 000 cm³ of **complemental air**. The maximum amount of air that you can move through your lungs is called the **vital capacity**. This is the total amount of air that moves through your lungs when you inhale and exhale as hard as you can. The vital capacity of an average person is about 4 500 cm³. A well trained athlete may have vital capacity of 6 500 cm³.

How much air do you breathe in during normal inspiration?

Procedure
Put a length of rubber tubing through the neck of a large container (a 5 dm³ polythene bottle is ideal) until it nearly reaches the bottom. Invert the bottle and place it under water in a laboratory sink (See Figure 1.4.6). Hold your nose and practise breathing in and out through your mouth. Breathe in air from the inverted bottle. Repeat four times. After breathing in, quickly remove the rubber tubing, insert a stopper in the bottle neck and remove the bottle. Measure the volume of water contained in the bottle using a measuring cylinder.

Repeat several times and take an average volume. Record the volume of water. What does this volume represent?

How much can you breathe out in a single breath?

Procedure
Set up the apparatus as shown in Figure 1.4.7. Practise deep breathing a few times, then blow

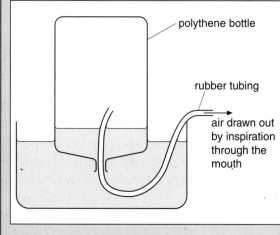

polythene bottle

rubber tubing

air drawn out
by inspiration
through the
mouth

Figure 1.4.6 Experimental apparatus

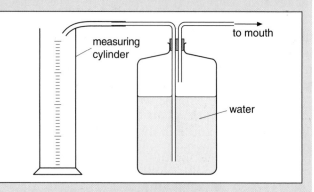

measuring
cylinder

to mouth

water

Figure 1.4.7 Experimental apparatus

LIFE PROCESSES

continued *out deeply through the mouth into the apparatus. Collect the displaced water in the largest available container (at least 5 dm³). Measure the volume of water collected using a measuring cylinder.*

Repeat several times and take your highest reading. This is your vital capacity. What is your vital capacity? Draw a bar chart for the vital capacities of all the students in your class. Compare this with the bar chart made for the breathing rates and exercise.

Our source of oxygen

The main gases that make up air are nitrogen, oxygen, and carbon dioxide. The gases have properties that are vital for life. These gases can diffuse through membranes. For instance, if there are different concentrations of oxygen on either side of a membrane, more oxgyen molecules will pass through the membranes from high to low concentration until the oxygen concentration is the same on both sides. Then the molecules will move at the same rate in both directions.

Gases in air can also dissolve in water. This is why oxygen and carbon dioxide can be carried by the blood. The solubility of the gases depends on temperature. Warm water dissolves much less oxygen than cold water. As water warms up, so gases come out of solution as bubbles.

Table I The composition of air (excluding rare gases, e.g. argon, neon, radon)

Gas	Inspired	Expired
oxygen	21%	16%
nitrogen	78%	77%
carbon dioxide	0.04%	4%
water vapour	variable	saturated

Gas exchange

The **pulmonary artery** (see pages 40 and 45) carries **deoxygenated blood** to the lungs. There it branches into an extensive network of small capillaries which completely surround each air sac (Figure 1.4.8).

The air in the alveolus and the blood in the capillaries contain gases in different concentrations. Therefore diffusion occurs through the thin, moist membranes of both alveolus and the capillaries. Oxygen diffuses from the air into the blood and carbon dioxide diffuses from the blood into the air sac.

The transport of oxygen

Oxygen is not very soluble in the plasma of the blood. It is even less soluble at our body temperature of 37°C. The red cells contain **haemoglobin** which has such a chemical attraction for oxygen that blood can be up to 97% saturated with it.

oxygen + haemoglobin $\rightleftharpoons$ oxyhaemoglobin

When blood reaches tissues which have a low concentration (**partial pressure**) of oxygen, the oxyhaemoglobin breaks down. The oxygen diffuses into the tissue fluid and, from there, reaches the cells. The attraction of haemoglobin for oxygen decreases with increasing acidity. This is important because, during exercise, carbon dioxide which is an acid gas is produced by the active muscle cells. **Lactic acid** may also be formed if the supply of oxygen to the cells is not sufficient. The increase in acidity causes the haemoglobin to release more of its oxygen just when it is needed.

LIFE PROCESSES

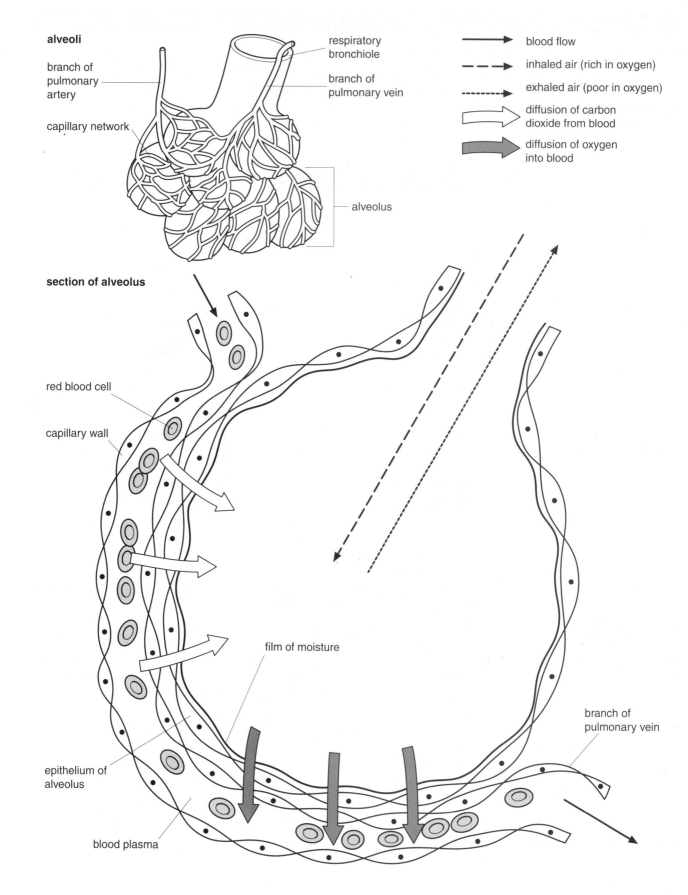

alveoli

branch of
pulmonary
artery

capillary network

respiratory
bronchiole

branch of
pulmonary vein

alveolus

blood flow

inhaled air (rich in oxygen)

exhaled air (poor in oxygen)

diffusion of carbon
dioxide from blood

diffusion of oxygen
into blood

section of alveolus

red blood cell

capillary wall

film of moisture

epithelium of
alveolus

blood plasma

branch of
pulmonary
vein

Figure 1.4.8 Gas exchange in an alveolus

The transport of carbon dioxide

Carbon dioxide is much more soluble than oxygen. It passes through membranes quickly and goes into the bloodstream. In the blood, only about 15% of the carbon dioxide is dissolved in the plasma. About 85% is carried as sodium hydrogen carbonate to the lungs where carbon dioxide is released from it and diffuses through the membranes of the capillaries and into the air sacs. From here, it leaves your body in the expired air.

Oxygen debt

During times of great muscular activity, the cells need more oxygen than the body can supply. The lungs cannot take in oxygen fast enough and the blood cannot deliver it fast enough. When this happens, the cells switch to **anaerobic respiration** (see page 8). This means that oxygen is not used and lactic acid collects in the tissues, causing a feeling of fatigue and even pain. The build up of lactic acid signals the brain's breathing centre to increase the breathing rate to supply the tissues with more oxygen.

If the heavy exercise continues, lactic acid keeps building up and causes an **oxygen debt**. It continues until the heavy exercise ends. Then, during a half-hour rest, some of the lactic acid is **oxidised** and some is converted to **glycogen**. The oxygen debt is paid and the body is ready for more exercise.

This athlete will have built up an oxygen debt in his muscles following the race

Carbon monoxide poisoning

Far too often, we read of people who have died as a result of carbon monoxide poisoning. Actually, the death is not caused by poisoning but by **tissue suffocation**. Carbon monoxide will not support life, yet it combines with haemoglobin 250 times more readily than oxygen does. As a result, the blood becomes loaded with carbon monoxide and its oxygen-combining power decreases. The tissues suffer from oxygen starvation and soon paralysis occurs followed by death.

Breathing at high altitudes

We live at the bottom of a large 'ocean' of air. As we move higher in it by climbing a mountain, the air pressure becomes less. Most of us have experienced 'popping' of our ears during an altitude change, perhaps in an aeroplane. It is due to there being a reduction in air pressure and our middle ear (see page 83) having to equalise the air pressure.

Pressure is important in determining how the oxygen combines with haemoglobin. This is why mountain climbers and aeroplane pilots have more difficulty in breathing as they move higher. Above 3 500m many people become tired easily. When an aeroplane approaches 6 000m, the pressure becomes so low that the pilot has difficulty in seeing and hearing.

This condition is a result of oxygen starvation of the tissues and is avoided by having an extra

Fighter pilots often wear oxygen masks at high altitude

These Peruvians are used to low levels of oxygen as they live at a high altitude in the Andes

supply of oxygen from an oxygen tank and mask. Passengers in airliners can fly at very high altitudes in the safety and comfort of pressurised cabins to counteract any natural pressure changes. In studying the problems of breathing at high altitudes, scientists observed Peruvian Indians.

These inhabitants of very high mountains in South America can carry out normal activities at altitudes at which most people would simply become exhausted. Scientists found that the Indians had greater than average lung capacity and an abnormally high red blood cell count. They had **adapted** to the high altitude. It was also noted that similar adaptations can take place in anybody who spends time at high altitudes. These people can exist in an atmosphere that has half the oxygen content that exists at sea level.

Summary

1 Breathing involves the exchange of gases between cells and their environment.

2 Breathing is a mechanical process that moves air in and out of the lungs due to expiration and inspiration.

3 During oxidation, foods break down, and energy is released.

4 In lower animals, individual cells are in direct contact with the environment, but in higher animals, blood is the transporting medium.

5 Atmospheric pressure and the concentration of gases in the air both play important roles in the diffusion of gases through membranes.

6 Solubility in water is an important property of gases that we exchange in our lungs.

LIFE PROCESSES

Questions for review

1 What are the differences between respiration and breathing?

2 Describe the mechanism that controls breathing rate.

3 What properties of gases aid the body in respiration?

4 Describe gas exchange in the lungs. Name the stuctures involved and explain why the exchange takes place.

5 How do pressure changes within the chest cavity cause inspiration and expiration?

6 Why is haemoglobin vital to the respiratory processes of the cells in our bodies?

7 How is oxygen carried from the lungs to the tissues?

8 How is carbon dioxide carried from the tissues to the lungs?

9 How do you build up an oxygen debt? How is it repaid?

10 Explain carbon monoxide poisoning.

Applying principles and concepts

1 What changes do you think would occur in the blood if you were to:
a) hold your breath for a period of time
b) breathe rapidly and deeply for a period of time?

2 Explain the decompression procedure used when divers come up from great depths.

3 What differences would you find in the blood of a person living at high altitudes compared to a person living at sea level?

4 Why is the fact that carbon dioxide is so soluble in water of vital importance to organisms?

LIFE PROCESSES

LIFE PROCESSES

Excretion

Learning Objectives

By the end of this chapter, you should be able to:

■ Identify the parts of the kidney
■ Understand that the kidney is an organ of excretion and regulation
■ Understand how a nephron works
■ Understand the function of a kidney machine

■ Identify the parts of the skin
■ Understand that the skin is an organ of excretion
■ Identify the functions of the skin

The removal of wastes of metabolism

Metabolism is the sum of the **breaking down processes (catabolism)** and the **building up processes (anabolism)** taking place in the body. During metabolism, many chemical reactions taking place in the body produce wastes which have to be removed or they would poison our organ systems.

When the body oxidises fuel during **respiration** it produces the waste products carbon dioxide and water. These are removed from the lungs, during **gaseous exchange** (see pages 57 and 58) and so the lungs have an excretory function.

One-celled animals excrete their wastes directly into the water-filled environment in which they live. Simple animals like jellyfish and sponges also do this, but the process is not so simple for organsisms made of millions of cells like ourselves. In humans, each cell discharges its waste materials into the tissue fluid which passes to the bloodstream. Then the blood carries the wastes to the excretory organs – the lungs, kidneys and skin.

Kidneys - the main excretory organs

The kidneys are bean-shaped organs, about the size of your clenched fist. They lie on either side of the spine, at the back of your abdominal cavity. Thick layers of **fat** cover and protect each kidney, and in order to see any details, you have to cut through a kidney lengthwise.

The firm outer part is the cortex, covered by a membrane, the capsule (Figure 1.5.2). The cortex makes up about a third of the kidney

tissue and is on the outside of the **medulla** which is filled with cone-shaped projections called **pyramids** pointing to a sac-like cavity, the **pelvis**. From here leads a long, narrow tube, the **ureter**, to the **bladder**. This is all you will see by cutting through the kidney and looking at the surface with a hand lens. In order to see further detail, it will be necessary to view thin sections of kidney tissue with a microscope.

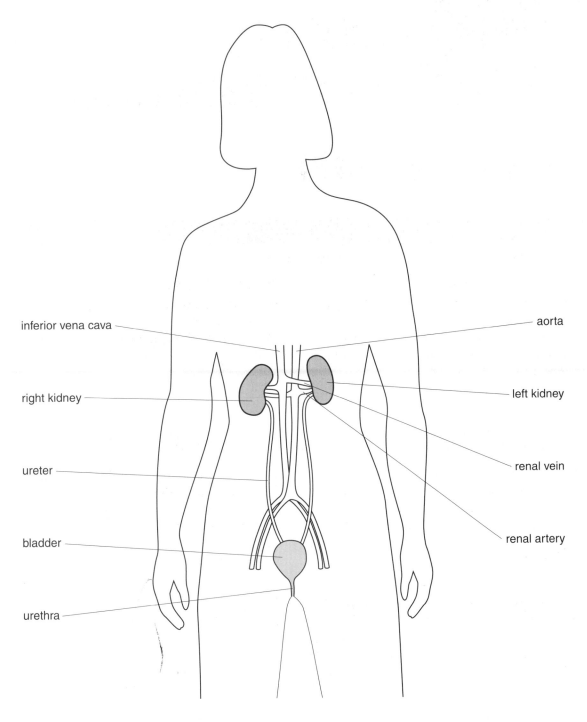

Figure 1.5.1 The urinary system

Each kidney has about a million tiny filters, all working together and separating large molecules from small ones as blood is forced through them under pressure. These filters are called **nephrons** (Figure 1.5.3) and control the chemical composition of the blood. They are responsible for the role of **regulation** performed by the kidneys. Together they measure more than 280 kilometres.

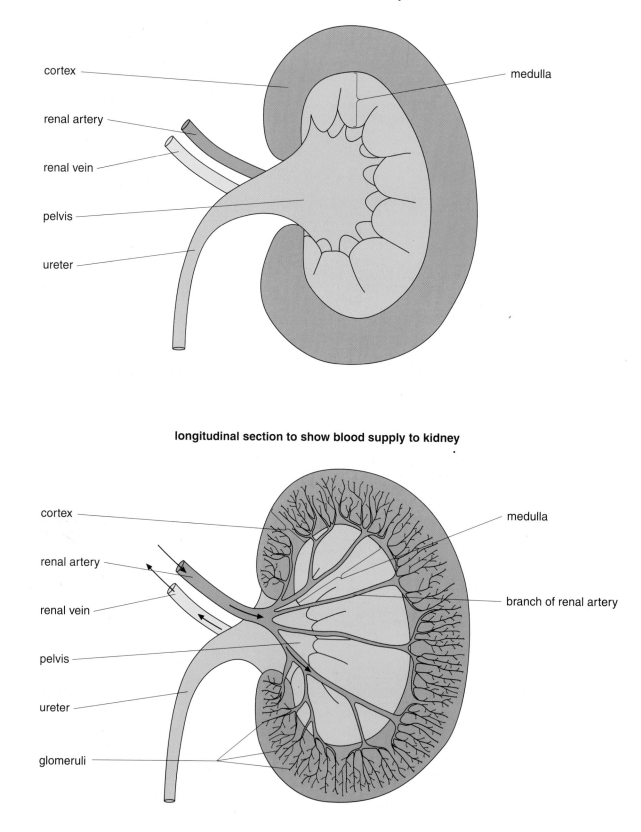

longitudinal section of kidney

cortex

renal artery

renal vein

pelvis

ureter

medulla

longitudinal section to show blood supply to kidney

cortex

renal artery

renal vein

pelvis

ureter

glomeruli

medulla

branch of renal artery

Figure 1.5.2 A section through the kidney

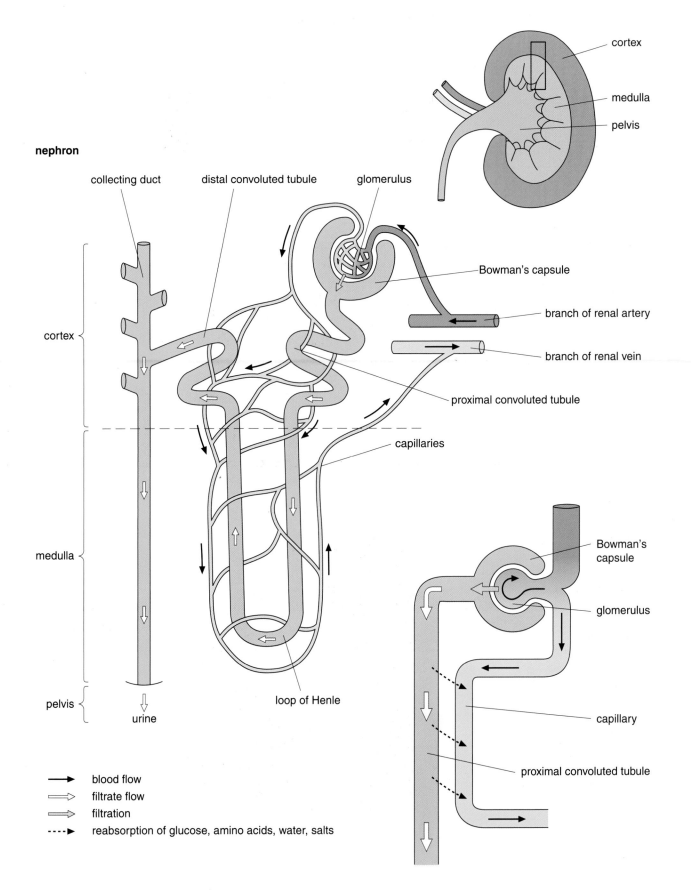

nephron

cortex

medulla

pelvis

collecting duct

distal convoluted tubule

glomerulus

Bowman's capsule

branch of renal artery

branch of renal vein

proximal convoluted tubule

capillaries

loop of Henle

urine

Bowman's capsule

glomerulus

capillary

proximal convoluted tubule

→ blood flow

⟹ filtrate flow

⟹ filtration

----► reabsorption of glucose, amino acids, water, salts

Figure 1.5.3 The structure of a nephron

LIFE PROCESSES

How does the nephron work?

Blood enters each kidney through a large **renal artery** which is a branch of the aorta. In the kidneys, each artery branches and rebranches into a mass of arterioles which eventually end in knots of blood capillaries called **glomeruli**. Each glomerulus fills a cup-like **Bowman's capsule** and, together, these form the actual filters of the kidneys.

Kidneys deal mainly with removal of **urea** which is formed in the liver (see page 25). There are two stages to removing urea and excess water and minerals from the blood: **filtration** and **reabsorption**.

The pressure of the blood is very high in the glomeruli. Note that, in Figure 1.5.3, the diameter of the in-going arteriole is greater than that of the out-going arteriole. Physically, this increases the blood pressure, and remember that the heart is pumping blood through the aorta directly into the renal arteries. Water, urea, glucose, amino acids and minerals are forced by this **ultrafiltration** through the capillary walls of the glomerulus into the surrounding Bowman's capsule. The liquid that passes through (the **filtrate**) is like blood plasma without blood proteins or cells. This filtrate, however, contains materials which are essential for the correct balance of the body's fluids. We cannot afford to lose glucose as it is our fuel. Neither can we afford to lose water if we are in danger of dehydration, or minerals if they are not in excess. So this is corrected in the second stage of the process.

After the filtrate leaves the Bowman's capsule it moves through the **tubule** which is surrounded by a network of capillaries. These blood vessels reabsorb glucose, minerals and amino acids by **active transport**. **Osmosis** is responsible for the reabsorption of water (see page 5). For every litre of filtrate, up to 990 cm³ may be reabsorbed. What is left passes through the tubules into the pelvis of the kidney as urine.

The regulation of water reabsorption is controlled by a hormone called **anti-diuretic hormone (ADH)**. There is no direct link between our alimentary canal and our urinary system, yet the more we drink, the more we urinate. The explanation is through the production of ADH when we need it. At times

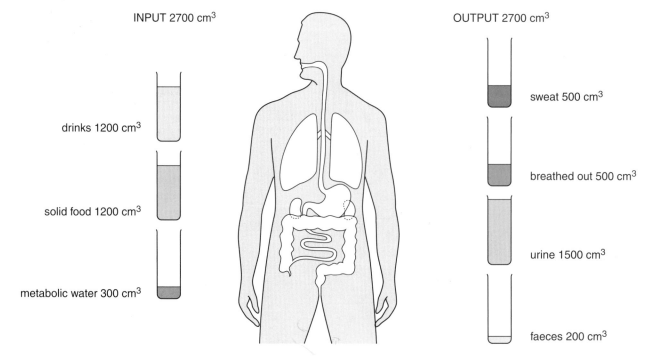

INPUT 2700 cm³

OUTPUT 2700 cm³

drinks 1200 cm³

solid food 1200 cm³

metabolic water 300 cm³

sweat 500 cm³

breathed out 500 cm³

urine 1500 cm³

faeces 200 cm³

Figure 1.5.4 Daily input and output of water in the body

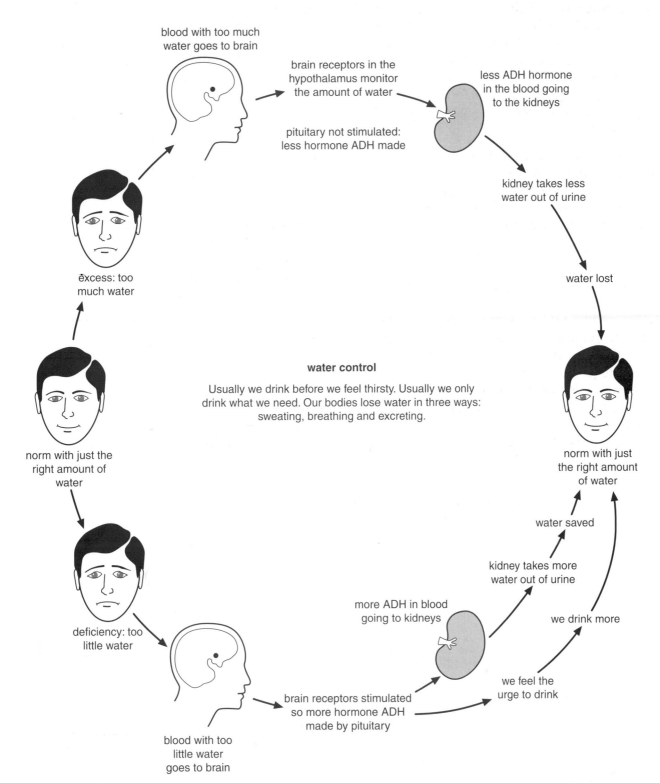

blood with too much
water goes to brain

brain receptors in the
hypothalamus monitor
the amount of water

less ADH hormone
in the blood going
to the kidneys

pituitary not stimulated:
less hormone ADH made

kidney takes less
water out of urine

excess: too
much water

water lost

water control

Usually we drink before we feel thirsty. Usually we only
drink what we need. Our bodies lose water in three ways:
sweating, breathing and excreting.

norm with just
the right amount
of water

norm with just the
right amount of
water

water saved

kidney takes more
water out of urine

more ADH in blood
going to kidneys

we drink more

deficiency: too
little water

we feel the
urge to drink

brain receptors stimulated
so more hormone ADH
made by pituitary

blood with too
little water
goes to brain

Figure 1.5.5 Water regulation

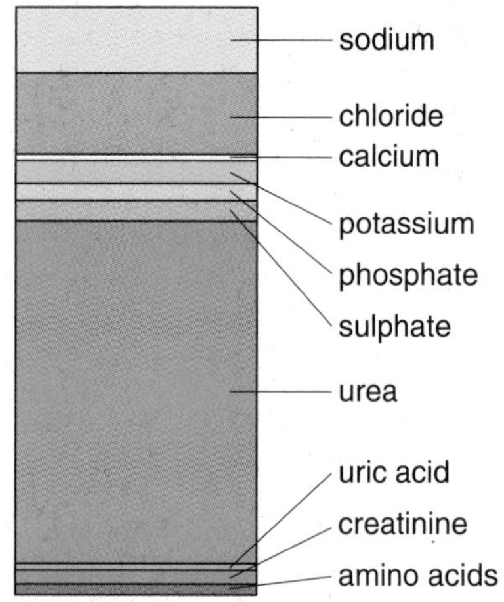

Figure 1.5.6 The composition of urine

sodium

chloride
calcium

potassium
phosphate
sulphate

urea

uric acid
creatinine
amino acids

of water shortage, when we are thirsty, there is very little water absorbed into the blood supply of the alimentary canal. Consequently, the blood becomes more 'concentrated' and, when it reaches a gland in the brain called the **hypothalamus**, this gland responds to the high concentration by secreting ADH into the blood. The blood carries the ADH to the kidneys and it causes changes in the cell membranes so that as

much water as possible is reabsorbed into the blood capillaries surrounding the tubules. At times like this, we excrete small volumes of very concentrated urine.

After drinking a lot of fluids, the concentration of our blood is decreased and the secretion of ADH by the hypothalamus is switched off. There is now nothing to cause the tubules to reabsorb more water so we excrete large volumes of very dilute urine.

The urine is made up mainly of water, urea, and mineral salts.

It passes from the pelvis of the kidney to the ureters which pass the urine to the bladder by **peristalsis**. The filtered blood is now the purest in the body and leaves the kidneys in the **renal veins**. These lead to the inferior vena cava, which in turn, carries the blood to the right atrium of the heart. The bladder is a muscular sac which can expand to hold the urine until its contractions push urine through the **urethra**.

Your kidneys have tremendous reserve power. If one has to be removed, the other will grow bigger and take over the job of both.

Living with kidney failure

Even though it is possible to live a normal life with one kidney, there are some people who have two kidneys which will not function properly. Where this is the case, the problem

has to be remedied with a kidney transplant or with a kidney machine. Some potential hazards of kidney transplants are discussed on page 226.

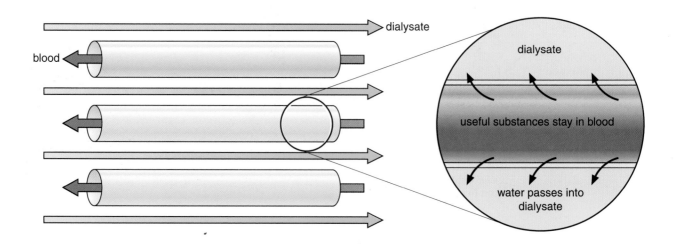

blood

dialysate

dialysate

useful substances stay in blood

water passes into dialysate

Figure 1.5.7 Kidney dialysis

Kidney machines work like real kidneys in some ways. The person's blood flows on one side of a very thin dialysing membrane. On the other side, a solution flows which resembles plasma without urea. This is called the **dialysate**. Waste products from the blood pass across the membrane into the dialysate in a process called **dialysis**. The dialysate has sugar and salts dissolved in it just like plasma. If this were not the case, useful substances, such as sugar and salts, would also pass out of the blood, together with urea by **diffusion**.

A person must be connected to the machine for about ten hours, two or three times a week. Blood is taken from a large vein which is surgically connected to an artery, usually in the arm or leg. It flows into the machine, becomes filtered by dialysis, and is pumped back into the same vein. The blood has to pass many times through the machine for complete filtration. This is why the person needs to be connected to the machine for so long.

Some people have their own kidney machine at home. Others go into a hospital each time they need dialysis. Although kidney machines undoubtedly save many lives, they can bring problems. Life can be restricted because dialysis

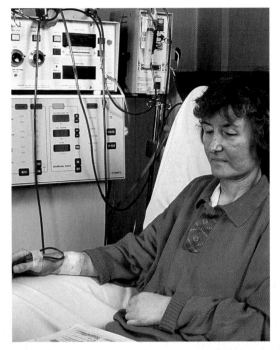

This woman has to spend many hours in hospital attached to a dialysis machine

takes so long, each week of each year. Anaemia and infections can be side effects. Also, a careful diet has to be followed to restrict foods that would contain many mineral salts or produce a lot of urea.

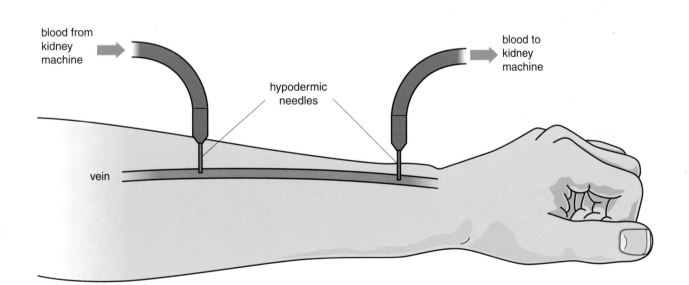

Figure 1.5.8 Connecting a patient to a kidney machine

What does a kidney filter out of the blood?

You are given three test tubes. You are asked to test the contents. In one there is a mixture of egg albumen, distilled water, glucose and urea. In another there is a mixture of distilled water, glucose and urea. In another there is a mixture of distilled water and urea. You do not know which is which. Carry out tests using the following materials to find which mixture represents:

a) Plasma from the renal artery

b) The filtrate from a kidney tubule

c) Urine.

Materials to be used

Clinistix for glucose – Changes to yellow. (An alternative is Benedict's Reagent, see page 19.)

Albustix for protein – changes to red. (An alternative is Biuret Reagent, see page 19.)

Urease – acts on urea to form an alkali (ammonium carbonate)

Indicator paper.

Explain the results by relating your observation to the way in which a kidney functions.

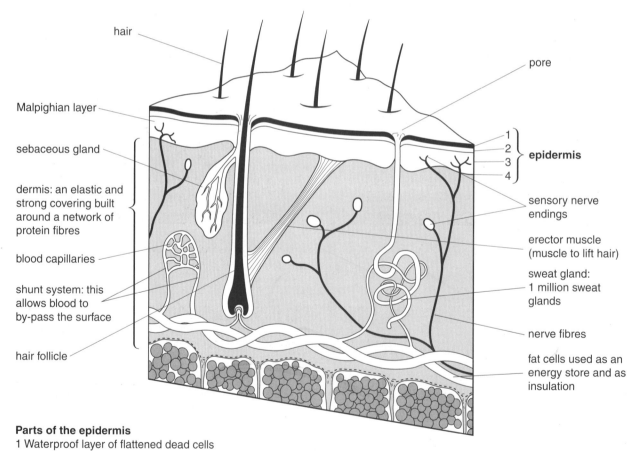

hair

pore

Malpighian layer

sebaceous gland

dermis: an elastic and strong covering built around a network of protein fibres

blood capillaries

shunt system: this allows blood to by-pass the surface

hair follicle

1
2
3
4

epidermis

sensory nerve endings

erector muscle (muscle to lift hair)

sweat gland: 1 million sweat glands

nerve fibres

fat cells used as an energy store and as insulation

Parts of the epidermis
1 Waterproof layer of flattened dead cells
2 Thinner layer of non-dividing living cells
3 Pigmented layer: protective, absorbs harmful ultraviolet light
4 Inner layer of dividing cells to replace those worn away

Figure 1.5.9 Skin structure

The skin as an excretory organ

The skin excretes water, minerals and some urea when we produce sweat. However, this fluid has a useful role because it helps to regulate your body temperature.

When liquid water is changed to water vapour by evaporation, heat is used for this change of matter from one state to another. So, as water in sweat evaporates from the body surface, heat is withdrawn from the outer tissues. The skin acts like an automatic radiator. It is richly supplied with warm blood and so heat is carried to the surface in this way. At the same time, production of sweat increases and bathes the skin. This increases the rate of evaporation and the amount of heat loss.

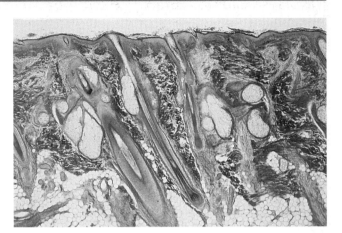

A photomicrograph of a skin section ×18

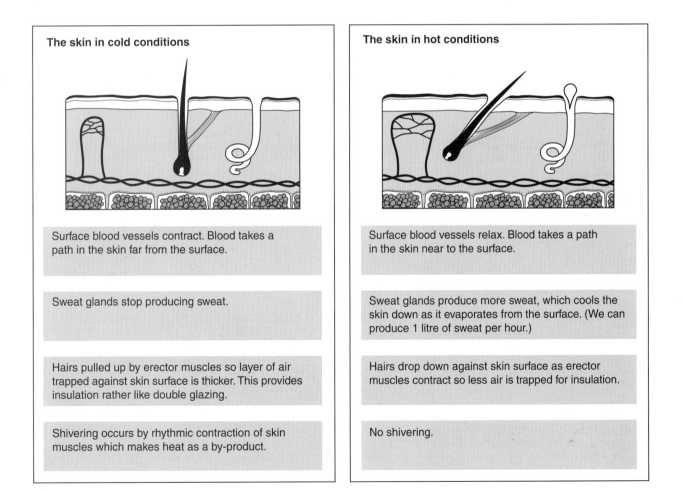

The skin in cold conditions

Surface blood vessels contract. Blood takes a path in the skin far from the surface.

Sweat glands stop producing sweat.

Hairs pulled up by erector muscles so layer of air trapped against skin surface is thicker. This provides insulation rather like double glazing.

Shivering occurs by rhythmic contraction of skin muscles which makes heat as a by-product.

The skin in hot conditions

Surface blood vessels relax. Blood takes a path in the skin near to the surface.

Sweat glands produce more sweat, which cools the skin down as it evaporates from the surface. (We can produce 1 litre of sweat per hour.)

Hairs drop down against skin surface as erector muscles contract so less air is trapped for insulation.

No shivering.

Figure 1.5.10 Temperature regulation

Excretion

Summary

1 Various wastes result from the metabolism of protein, carbohydrate and fat.

2 They are removed from the body with the help of the kidneys, skin, lungs and liver.

3 The kidneys are the most important excretory organs.

4 They filter practically all the urea from the blood.

5 The skin also excretes wastes in sweat.

6 Sweat helps to control your body temperature.

Questions for review

1 How do the kidneys regulate the contents of the blood?

2 What are the differences between the filtrate in the Bowman's capsule and urine?

3 What are the differences between the contents of the blood plasma of the renal artery and that of the renal vein?

Applying principles and concepts

1 Why is increased salt intake recommended in hot weather?

2 How do the kidneys help in maintaining the water balance of the body?

LIFE PROCESSES

LIFE PROCESSES

Body controls

Learning Objectives

By the end of this chapter you should be able to:

- List the main parts of the central nervous system
- Describe the basic structure of a nerve cell
- Name the main parts of the brain
- Describe the functions of each major part of the brain
- Identify the main areas of the spinal cord
- Understand a reflex arc
- Describe the major body senses
- Demonstrate a knowledge of the eye and ear

The nervous system

A nervous system is used by animals to make them aware of changes in their surroundings (**stimuli**) and to help co-ordinate a **reaction** to stimuli in such a way that it is to the animal's advantage.

The nervous system is divided into three main parts. These are:

1 the brain and spinal cord, making up the **central nervous system**

2 the **peripheral nervous system**, consisting of nerves connencting the central nervous system with all organs of the body

3 the **autonomic nervous system**, which regulates all the automatic actions of organs that carry out vital activities such as the heart beat, breathing rate, peristalsis, etc.

Nerve cells are called **neurons** (Figure 1.6.1) and each acts as a link in the nervous system because of its unique structure.

Each neuron has a star-shaped **cell body** containing the bulk of the cytoplasm and the nucleus. Thread-like projections, called **nerve fibres**, extend from the cell body and are used to conduct messages, called **impulses**, to their destinations.

We react to changes in our environment through our **senses**. When you stub your toe, the impulse travels to your central nervous system and is relayed very quickly back to muscles in your foot. These contract to move your foot away. In the meantime, your brain has registered the feeling of pain and you are aware of the problem.

Nerve cell fibres run in bundles called **nerves**. They are something like an electrical cable, made up of insulated wires bound together. In the peripheral nervous system any nerves that carry impulses that activate muscles are called **motor nerves**. **Sensory nerves** carry impulses away from sense organs towards the central nervous system. Nerve fibres of one neuron never really touch those of another because there is always a tiny space between them. The spaces are called **synapses** and have to be

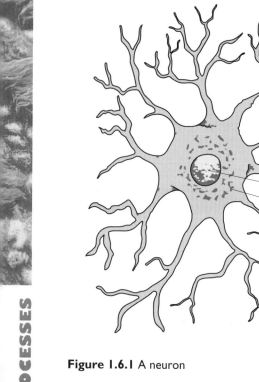

Figure 1.6.1 A neuron

crossed by impulses before any message can be sent from one neuron to another. Furthermore, an impulse never travels from one motor neuron to another or from one sensory neuron to another. Impulses bridge the synapse with the help of fast-acting chemical reactions in **transmitter substances** produced at the endings of neurons.

The brain

The part of the brain called the **cerebrum** is proportionally larger in humans than in any other animal (Figure 1.6.2). It consists of two halves, the **cerebral hemispheres**, which have an outer surface called the **cortex** that is deeply folded in irregular wrinkles and furrows. These greatly increase the surface area so that countless numbers of neurons can develop in this region. Indeed, the vast numbers of cell bodies of these neurons give this tissue its name of **grey matter**. The cerebrum below the cortex is called the **white matter** because it is formed of masses of nerve fibres surrounded by insulating white fatty sheaths of **myelin**. These fibres extend from the neurons of the cortex to other parts of the body.

Specific regions of the cerebrum control specific activities. Some areas of the cerebral cortex are called **motor areas** and control voluntary movements. Also, there are **sensory areas** that interpret sensations. The cerebrum is responsible for storing information and for processes that we associate with complex activities, thinking and intelligence.

The **cerebellum** lies below the back of the cerebrum and both work together in controlling muscular activity. Nerve impulses do not originate in the cerebellum and you cannot control its activities. It is the cerbellum that coordinates the motor impulses sent from the cerebrum. Without the help of the cerebellum, the cerebrum's impulses would produce unco-ordinated motions.

Another function of the cerebellum involves the maintenance of balance. Impulses from your eyes and inner ears (see page 82) inform the cerebellum of your position relative to your surroundings. Then the cerebellum controls the muscular contractions necessary to maintain balance. It can do this by automatically maintaining tone in muscles which keeps muscles in a state of partial contraction.

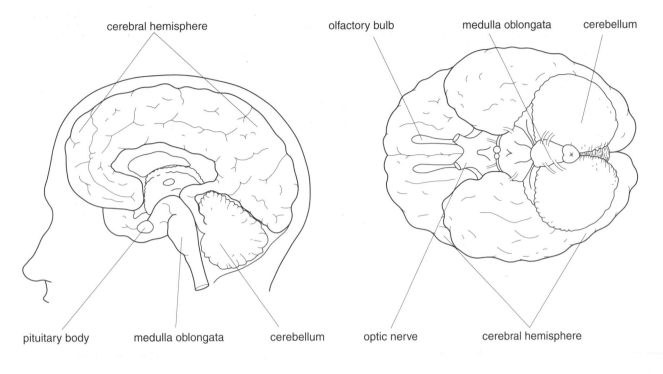

Figure 1.6.2 The human brain

The **brain stem (medulla oblongata)** is an enlargement at the base of the brain. This is where nerve fibres from the cerebrum and the cerebellum collect before leaving the brain. It controls the activity of the internal organs and also contains the **respiratory control centre** (see page 75). All automatic activities are controlled here, e.g. the heart action, peristalsis, glandular secretion, diaphragm action, and dilation of blood vessels.

The spinal cord and spinal nerves

The spinal cord extends down from the brain stem, passing through the protection of the vertebrae along the whole length of the spine. Unlike the brain, the positions of the grey and white matter are reversed. In transverse section, the grey matter in the spinal cord takes on the shape of a butterfly with outspread wings, as you can see in the photo on page 77.

Thirty-one pairs of spinal nerves branch off the nerve cord, passing out between the vertebrae, and branch to make up most of the peripheral nervous system. Each spinal nerve divides just outside the cord. The sensory fibres that carry impulses from the sense organs to the cord go through the dorsal root and enter the grey matter. Cell bodies of the sensory fibres form a swelling called the **dorsal root ganglion** on this branch of the spinal nerve.

The other branch at this junction is the ventral root and carries the motor fibres from the grey

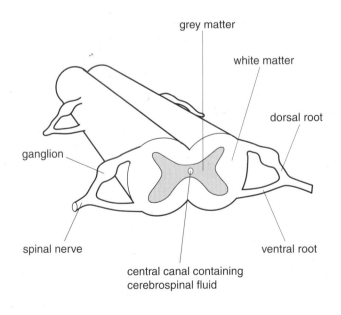

Figure 1.6.3 The structure of the spinal cord

LIFE PROCESSES

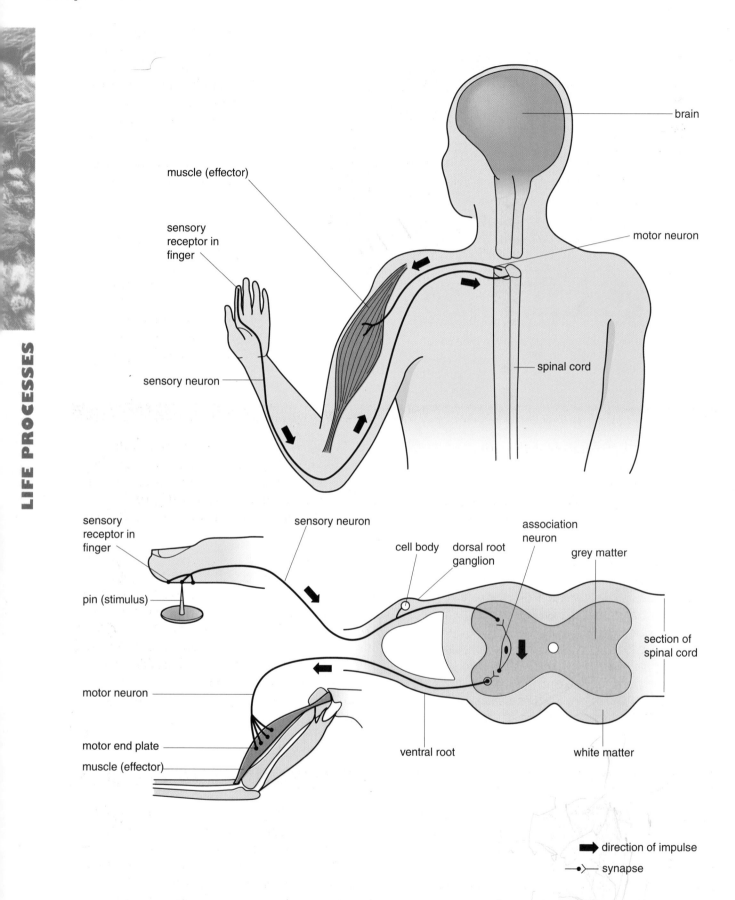

muscle (effector)

sensory
receptor in
finger

sensory neuron

brain

motor neuron

spinal cord

sensory
receptor in
finger

sensory neuron

cell body

dorsal root
ganglion

association
neuron

grey matter

pin (stimulus)

section of
spinal cord

motor neuron

motor end plate

muscle (effector)

ventral root

white matter

direction of impulse

synapse

Figure 1.6.4 A reflex arc

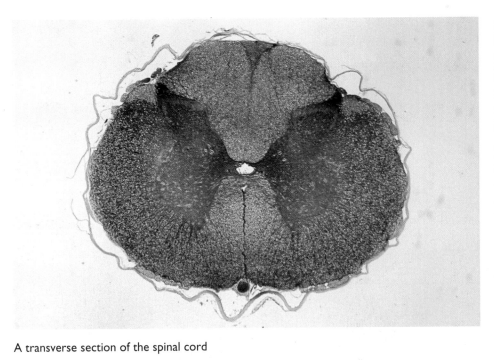

A transverse section of the spinal cord

matter where their cell bodies are located. The motor fibres of this branch carry impulses from the spinal cord to muscles or glands of the body.

If your spinal cord were cut, serious problems would result. All the parts of your body controlled by nerves that leave the cord below the point of the cut would be totally **paralysed**. In addition you would lose all sense of feeling in many areas below the damaged spinal cord.

Reflex actions

The simplest kind of nervous reaction in humans is called a reflex action. It can involve only two or three nerves which link a **receptor** (sense organ) to an **effector** (muscle or gland) via the central nervous system (spinal cord or the brain) (Figure 1.6.4).

It is a rapid, automatic reaction which does not involve conscious effort and may not involve the brain. The knee jerk is a good example of a simple reflex action. Sit on the edge of a table and let you knee swing freely. Then tap just below the knee cap with a narrow object. Your lower leg will jerk upward because the tap stimulates a sensory neuron in the lower leg. An impulse travels along the sensory neuron to an intermediate neuron in the centre of the

spinal cord, via a synapse. The **intermediate neuron** stimulates a motor neuron via another synapse and the impulse travels to muscles in the leg which contract, causing the jerking movement. The entire reflex takes a spilt second (Figure 1.6.5).

Reflex actions have a protective function. For example, when you touch a hot object, your hand jerks away almost instantly. The reflex is complete even before your brain registers the pain. If the muscle response were delayed until the pain impulse was complete and interpreted, the effects of the burn would be much greater. Other reflexes include sneezing, coughing, blinking, and the constriction or dilation of the pupil in the eye.

The skin as a sense organ

Specialised parts of sensory neurons make up the receptors found in the skin. They can be bare nerve endings or are collected into one or more cells. Each receptor is suited to receive only one type of stimulus which triggers an impulse to pass along a sensory neuron to the central nervous system.

There are five distinct kinds of receptors in the skin. They are specialised to respond to the stimuli of either touch, pressure, pain, heat or cold. The pain receptor, for example, is a bare ending of a neuron. If the stimulus is strong enough, a pain receptor will react to heat, mechanical, electrical, or chemical stimuli.

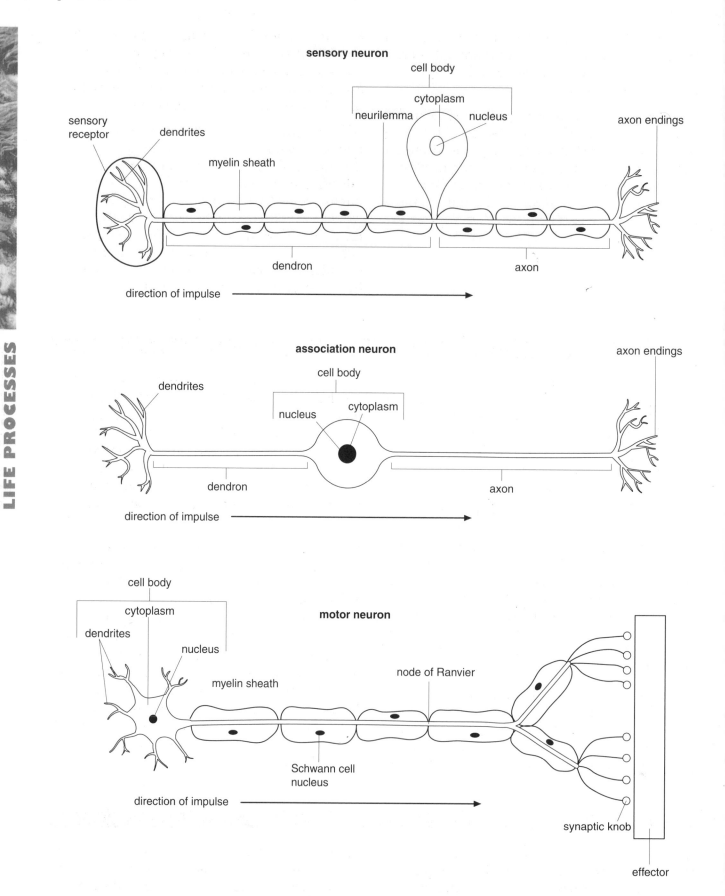

Figure 1.6.5 Neurons involved in reflex actions

The sensation of pain is protective and signals a threat of injury to the body. Therefore, it is an advantage for pain receptors to be located throughout the skin.

The sensory nerves lie at different depths in the skin. For example, if you move the point of a pencil over your skin very lightly, you stimulate only the nerves of touch because the receptors for touch are close to the surface of the skin. The fingertips, the forehead, and the tip of the tongue contain many receptors that respond to touch.

Receptors that respond to pressure lie deeper in the skin. If you press the pencil point against the skin, you feel both pressure and touch. Since the nerves are deeper, a pressure stimulus must be stronger than a touch stimulus.

Heat and cold stimulate different receptors and this is an interesting protective adaptation of the body. A feeling of cold results from a lowering in heat energy. If both great heat and intense cold stimulated a single receptor, we would be unable to tell the difference between the two, and we would be unable to react to either. However, since some receptors are stimulated by heat and others by the absence of it, we can react to both conditions.

The sense of taste

Nearly all animals prefer some foods to others because they can distinguish different chemicals. Like other senses, taste results from the stimulation of certain nerve endings. In this case, the stimulation is chemical and, in humans, the nerve endings for taste are located in flask-shaped **taste buds** (Figure 1.6.6) on the tongue.

They lie on the front of the tongue, along its sides, and near the back. Bits of food, mixed with saliva, enter the taste buds through little pores at the tops and stimulate hair-like nerve endings. The message carried to the brain from these nerve endings is interpreted as the sense of taste. Our sense of taste like our sense of smell is not very well developed. In fact, we can taste only four basic flavours: sour, sweet, salty, and bitter.

The taste buds for each of these flavours are located in different parts of the tongue. Those for sweetness are on the tip of the tongue. That is why chocolate tastes sweeter when you lick it than when you chew it. Salt sensitive buds are also on the tongue's tip. Those for sour flavours lie along the sides of the tongue, and those for bitterness lie on the back of the tongue. That is why, if you eat something both sweet and bitter, you taste the sweetness before the bitterness. Some foods, like pepper and other spices, have no distinct flavour. They taste the way they do because they irritate the entire tongue, causing a burning sensation.

Much of the sensation we call taste is really smell. When you chew some onion or apple, the vapours enter the inner openings of the nose. There, they reach the nerve endings for smell, and you can tell them apart. You may have noticed the loss of what you thought was taste when you had a cold. When your nose is blocked up with mucus, few food vapours can get to the nerve endings for smell. That is why food does not taste very good when you have a head cold. In fact, under these conditions, the apple and the onion may even have the same sweet flavour.

The sense of smell

Like taste, smell results from chemical stimulation of certain nerve endings (Figure 1.6.7). The difference is that, in smell, the chemical stimulator is in the form of gases which dissolve in mucus. Then they stimulate these nerve endings and cause impulses to go to the central nervous system where the impulses are interpreted as smell. However, if the smell receptors are exposed to a particular odour for a long time, they stop reacting to it but will

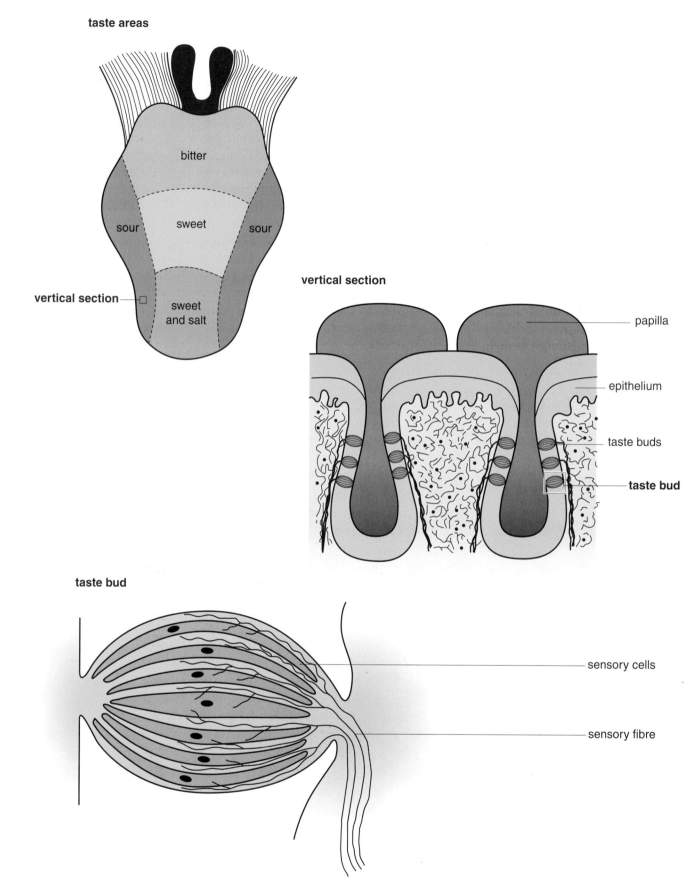

taste areas

bitter

sour

sweet

sour

vertical section

sweet
and salt

vertical section

papilla

epithelium

taste buds

taste bud

taste bud

sensory cells

sensory fibre

Figure 1.6.6 Taste buds

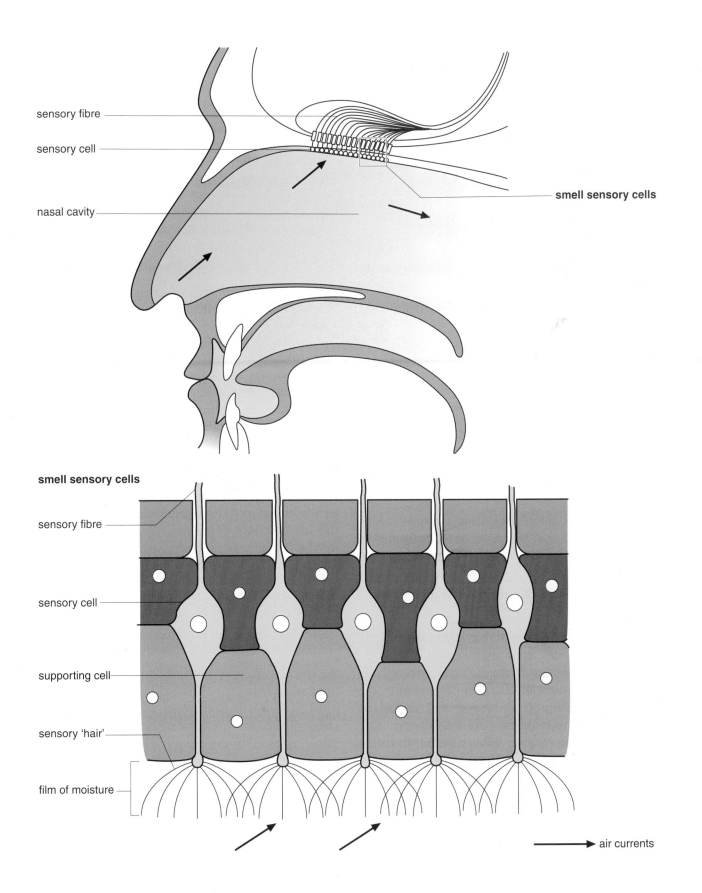

sensory fibre

sensory cell

nasal cavity

smell sensory cells

smell sensory cells

sensory fibre

sensory cell

supporting cell

sensory 'hair'

film of moisture

air currents

Figure 1.6.7 The sense of smell

respond to other odours. If you go into a laboratory where a particularly smelly chemical is being made, the smell may be over-powering to start with but you soon fail to notice it. The part of the brain responsible for interpreting the impulse has become fatigued and does not respond.

The sense of hearing

Like all mammalian ears, human ears are very complex organs, as can be seen in Figures 1.6.8 and 1.6.9.

Note the Eustachian tube which connects the middle ear with the throat. This connection equalises pressure in the middle ear with that of the outer atmosphere. When it becomes blocked by mucus from a cold, the inner and outer pressures do not equalise and, if the pressure difference becomes great enough, the eardrum may burst. For this reason professional divers and aeroplane pilots do not work when they have colds. The outside pressure increases during a dive and with a blocked Eustachian tube, the middle ear pressure would not be equalised. The difference might burst the eardrum. With the pilot, the situation would be reversed. The pressure would be less outside the middle ear than inside. That is why all commercial airliners have pressurised cabins.

We are able to hear because all noises are caused by vibrations. When an object vibrates in the air, it mechanically moves the air molecules and some of them are squeezed together or **compressed**. Others are spread apart, or **rarefied**. The regular pattern that is produced by any vibrating object in the air, or any other medium, is called a **sound wave**. These pass more effectively through liquids and solids than they do through air because the molecules are more closely packed together in the more dense media.

When sound waves reach the ear, they pass through the **auditory canal** and to the **eardrum**. The sound waves start the eardrum vibrating in the same patteren as the particular wave. The vibrating eardrum then starts the **hammer (malleus)**, the **anvil (incus)** and the **stirrup (stapes)** to vibrate. These, in turn, cause the **oval window** to vibrate which sets up vibration in the fluid inside the **cochlea**. The vibrations of the fluid stimulate the nerve endings that line the cochlea which in turn transmit impulses in the **auditory nerve** to the cerebrum. The specific pattern of the impulses is determined by the pattern of the sound wave that started the whole chain of events. The cerebrum picks up the impulses and translates them into a perception of sound.

The sense of balance

The **semicircular canals** are the structures necessary to our sense of balance. The canals contain many receptors and lie at right angles to each other on three different planes (Figure 1.6.10). They also contain a fluid similar to the fluid inside the cochlea. When your head changes position, this fluid rocks and stimulates the receptors. With this stimulation, the receptors start impulses that go through a branch of the auditory nerve to the cerebellum. The brain is then made aware of changes in head position. The canals lie in three different planes so that any changes in the head's position will rock the fluid in one or more directions.

People get dizzy when they spin around and then suddenly stop. The spinning forces the fluid to the ends of the canals. When the spinning stops, the fluid rushes back, causing the sensation of spinning in the opposite direction. The sensory conflict causes a sensation of dizziness or nausea. In some people, regular rhythmic motions produce unpleasant sensations that invole the whole body. When this occurs in a ship, aeroplane, or car, it is called **motion sickness**. If the semicircular canals become diseased, temporary or permanent dizziness may result.

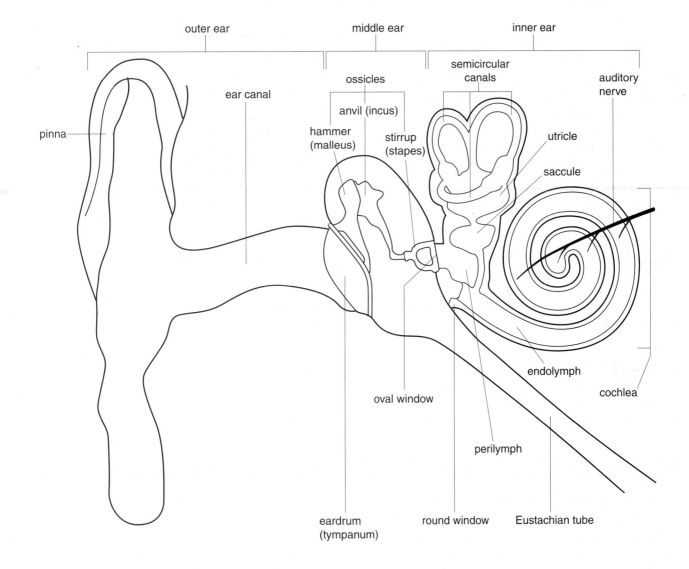

outer ear

middle ear

inner ear

semicircular
canals

ear canal

ossicles

auditory
nerve

anvil (incus)

pinna

hammer
(malleus)

stirrup
(stapes)

utricle

saccule

endolymph

cochlea

oval window

perilymph

eardrum
(tympanum)

round window

Eustachian tube

Figure 1.6.8 The ear

passage of sound waves through the ear

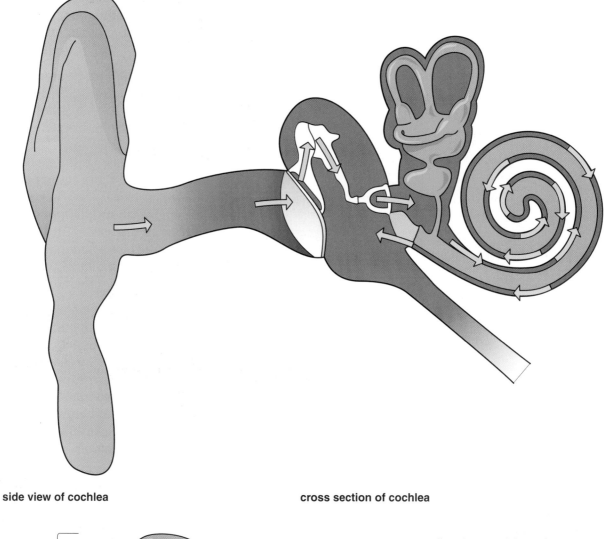

side view of cochlea

cochlea

auditory nerve

cross section

cross section of cochlea

perilymph tectorial membrane

branch of auditory nerve

perilymph

endolymph

basilar membrane sensory hair cell

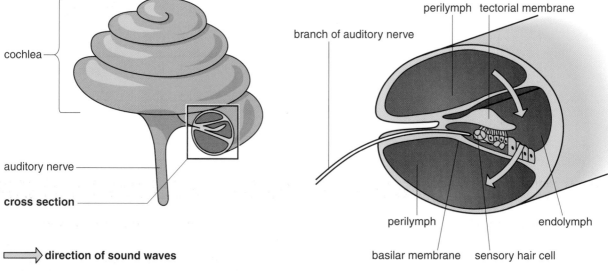

direction of sound waves

Figure 1.6.9 How we hear

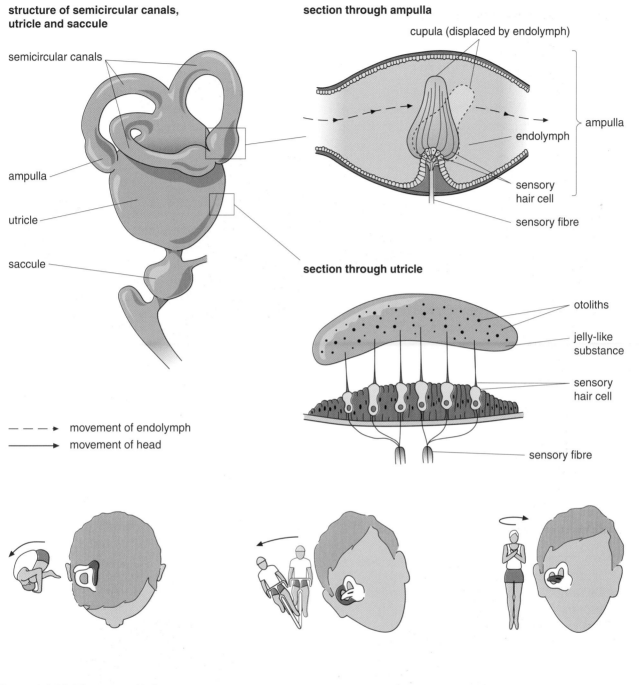

structure of semicircular canals, utricle and saccule

semicircular canals

ampulla

utricle

saccule

- - - ► movement of endolymph
——————► movement of head

section through ampulla

cupula (displaced by endolymph)

ampulla

endolymph

sensory hair cell

sensory fibre

section through utricle

otoliths

jelly-like substance

sensory hair cell

sensory fibre

LIFE PROCESSES

Figure 1.6.10 The sense of balance

The sense of sight

The function of almost all the structures of the eye is to focus light on the **retina** (Figure 1.6.11). The retina contains the receptors that are stimulated by light. The receptors are of two types, called **rods** and **cones**, and when they are stimulated, they start impulses which eventually pass through the **optic nerve**. There are no rods or cones at the spot where the optic nerve joins the retina. Therefore, there can be no vision at this point and so it is called the **blind spot**. The optic nerve from each eye leads to the vision centre of the brain in the cerebrum.

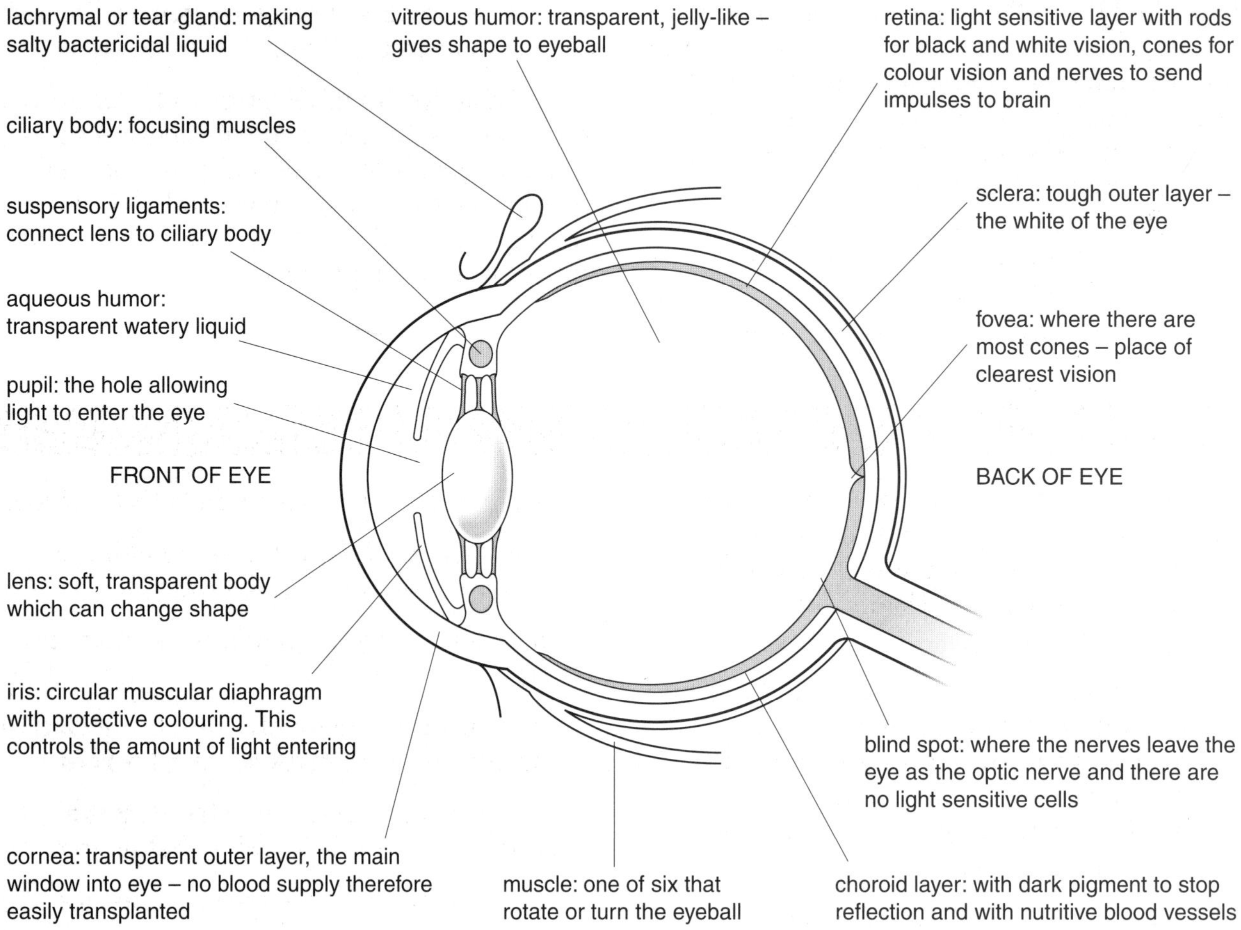

lachrymal or tear gland: making salty bactericidal liquid

ciliary body: focusing muscles

suspensory ligaments: connect lens to ciliary body

aqueous humor: transparent watery liquid

pupil: the hole allowing light to enter the eye

FRONT OF EYE

lens: soft, transparent body which can change shape

iris: circular muscular diaphragm with protective colouring. This controls the amount of light entering

cornea: transparent outer layer, the main window into eye – no blood supply therefore easily transplanted

vitreous humor: transparent, jelly-like – gives shape to eyeball

muscle: one of six that rotate or turn the eyeball

retina: light sensitive layer with rods for black and white vision, cones for colour vision and nerves to send impulses to brain

sclera: tough outer layer – the white of the eye

fovea: where there are most cones – place of clearest vision

BACK OF EYE

blind spot: where the nerves leave the eye as the optic nerve and there are no light sensitive cells

choroid layer: with dark pigment to stop reflection and with nutritive blood vessels

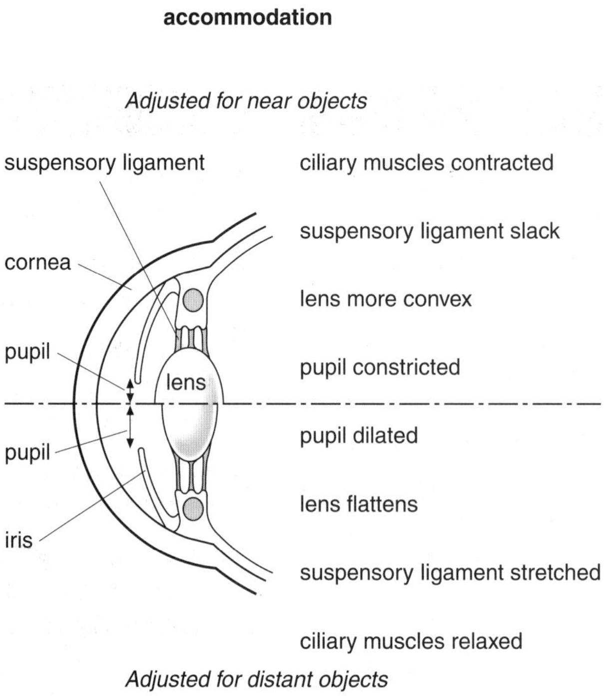

accommodation

Adjusted for near objects

suspensory ligament — ciliary muscles contracted

suspensory ligament slack

cornea — lens more convex

pupil — pupil constricted

lens

pupil — pupil dilated

lens flattens

iris — suspensory ligament stretched

ciliary muscles relaxed

Adjusted for distant objects

Figure 1.6.11 The eye

The focusing mechansims of the eye attempt to bring light rays to the most sensitive part of the retina called the **yellow spot** or **fovea centralis**. This part has more cones than other parts of the retina. The cones are more sensitive to bright light than are the rods and are responsible for colour vision. An image formed as a result of light rays focusing outside the fovea is less distinct. This explains why, if you focus your eyes directly on an object, you see it clearly, while surrounding objects tend to lack detail. In the dark, there is not enough light to stimulate the cones, but sometimes there is enough to stimulate the rods. However, rods cannot distinguish colours. Therefore, you do not see much colour in dim light.

The human eye contains fewer rods than many other animals' eyes and so our night vision is relatively poor. Cats, deer and owls see well at night because they have many rods. Most owls, however, lack cones in their eyes and therefore see less well during the day.

Summary

1 Humans have the most complex central nervous system of any animal. It is composed of the brain and spinal cord and communicates with all parts of the body via the nerves.

2 The cerebrum controls conscious activities and is the centre of intelligence. The cerebellum co-ordinates impulses and controls movement and balance. The medulla oblongata controls the automatic actions of the body.

3 Sensory nerves carry impulses to the central nervous system from sense receptor organs which include those of the:

skin, responsible for the sense of touch, pressure, pain, heat and cold

tongue and nose, responding to chemical stimuli

cochlea of the inner ear, which respond to vibrations in the form of sound waves

semicircular canals, which detect the position of your head when balancing.

rods and cones of the retina in the eye which respond to light.

Questions for review

1 Name the main divisions of the nervous system and state the functions of each.

2 Name the main parts of the brain and give brief functions of each.

3 Name the five sensations of the skin. In what ways are the receptors different?

4 Describe how sound waves in the air stimulate the receptors in the cochlea.

5 Why is our night vision relatively poor compared to the night vision of an owl?

Applying principles and concepts

1 Why are peripheral nerves that contain only axons considered to be motor nerves?

2 In what way is the autonomic nervous system really two systems?

3 Account for the fact that we think we distinguish more than the four tastes the tongue can actually detect.

4 How can an infection in the middle ear produce temporary deafness?

5 Describe the movements of the head that would be necessary to stimulate each semicircular canal separately.

LIFE PROCESSES

LIFE PROCESSES

Body regulators

Learning Objectives

By the end of this chapter, you should be able to:

- **Identify the endocrine glands**
- **Explain the function of each of the endocrine glands**

- **Explain how the endocrine glands work together in a balanced way.**

The endocrine system

Some glands, like the salivary glands, pour secretions into the digestive system through ducts or tubes. These are called **exocrine glands**. **Endocrine glands** are quite different because they are ductless and their secretions go directly into the bloodstream which carries them to all parts of the body. The secretions of ductless glands are called **hormones** and they regulate the activities of all the body processes.

The thyroid

The thyroid gland is in the neck, near the lower part of the larynx where it meets the trachea. It consists of two lobes connected by a narrow bridge. The hormone produced by the thyroid is called **thyroxine** and has the highest concentration of iodine found in any substance in the body. Purified thyroid extract from sheep is used for treating thyroid disturbances in humans. Although it is the least expensive of all commercial endocrine preparations, its supply can be increased by **genetic engineering**.

Thyroxine regulates certain metabolic processes, especially the ones related to growth and respiration within cells. If the thyroid gland is overactive, it produces a condition called **hyperthryroidism**. The rate of cellular respiration increases, the body temperature increases, and the heart rate goes up together with blood pressure.

Sweating, when the body should be cool, is a common symptom and the person gets very nervous and irritable. Some people's eyes bulge slightly, and they develop a staring expression. Surgery used to be the only treatment for this condition. However, an effective drug called **thiouracil** has been developed and doses of radioactive iodine can also be given as effective treatment. The radiation destroys some of the thyroid tissue.

If the thyroid gland is underactive, the symptoms are the opposite of those of hyperthyroidism. People with this condition are characteristically physically or mentally retarded. Their hearts enlarge and the rate of heartbeat slows down. The condition can be treated with thyroid extract.

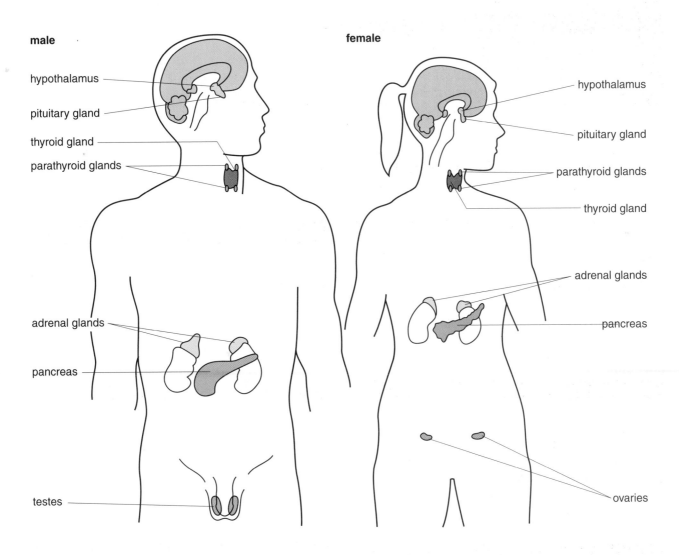

Figure 1.7.1 The endocrine system

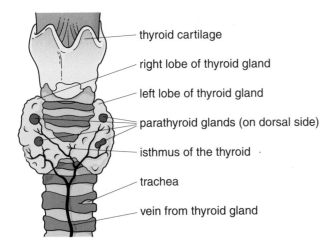

thyroid cartilage

right lobe of thyroid gland

left lobe of thyroid gland

parathyroid glands (on dorsal side)

isthmus of the thyroid

trachea

vein from thyroid gland

Sometimes the thyroid does not function properly during infancy. This results in **cretinism** which has symptoms of physical and mental retardation. Treatment must occur during critical stages of development to be successful. When iodine is deficient in the diet, the thyroid gland enlarges to form a goitre (see page 11).

People who eat seafood regularly rarely have this condition because of the iodine content. In areas where there is not much iodine in the soil, iodine can be added to the diet in additives to materials such as table salt.

Figure 1.7.2 The thyroid gland

Parathyroid glands

There are four parathyroid glands embedded in the back of the thyroid, two in each lobe. The parathyroids secrete **parathyroid hormone** which controls the body's use of calcium.

A constant, stable balance of calcium is vital to the body. Bone growth, muscle tone, blood clotting, and normal nervous activity depend on calcium balance.

The pituitary

The pituitary is a small gland, about the size of a pea, located at the base of the brain and used to be called the 'master gland' because its secretions affect the activity of all other endocrine glands (Figure 1.7.3). It is now known that other glands, in turn, affect the pituitary.

Two lobes make up the pituitary gland, anterior and posterior. The **anterior lobe** secretes several different hormones. One of these is **growth hormone** which controls the development of your skeleton – your whole body framework. Other secretions of this lobe are the **gonadotropic hormones**. They influence the development of the reproductive organs and also affect hormone secretions of ovaries and testes (see page 93). Other hormones from the anterior lobe stimulate milk production in the mammary glands, thryoxine in the thyroid, and hormones in the adrenal glands.

The world's tallest living woman, Sandy Allen (born 1955) with her 12-year-old brother. She's 2.31 m tall

The **posterior lobe** of the pituitary releases two hormones. One is oxytocin which helps regulate your blood pressure and stimulates your smooth muscles (see page 130). During childbirth, oxytocin is secreted in large amounts to make the uterus contract. The other hormone is **anti-diuretic hormone (ADH)** which controls water reabsorption in the kidneys (see page 66). A deficiency of this hormone causes **diabetes insipidus** which allows too much water to be lost in urine.

The most common disorder of the pituitary gland involves the growth hormone. If the pituitary secretes too much of this hormone during infancy, the bones and other tissue grow too fast and the person could grow to 2.5 m in height. This is **gigantism**. Deficiency of the growth hormone slows down growth to produce **midgets** which are not the same as dwarfs, who grow as a result of thyroid deficiency. These people are also mentally retarded.

The glands of emergency

On top of each kidney are your **adrenal glands**, also called supra renal glands (Figure 1.7.4).

These glands have an outer layer called the cortex, and an inner part called the **medulla**. Unlike the adrenal medulla, the adrenal cortex

is essential for life. The cortex secretes hormones called **corticoids** which control carbohydrate, fat and protein metabolism. They also affect the balance of salt and water in your body. Other hormones produced here control

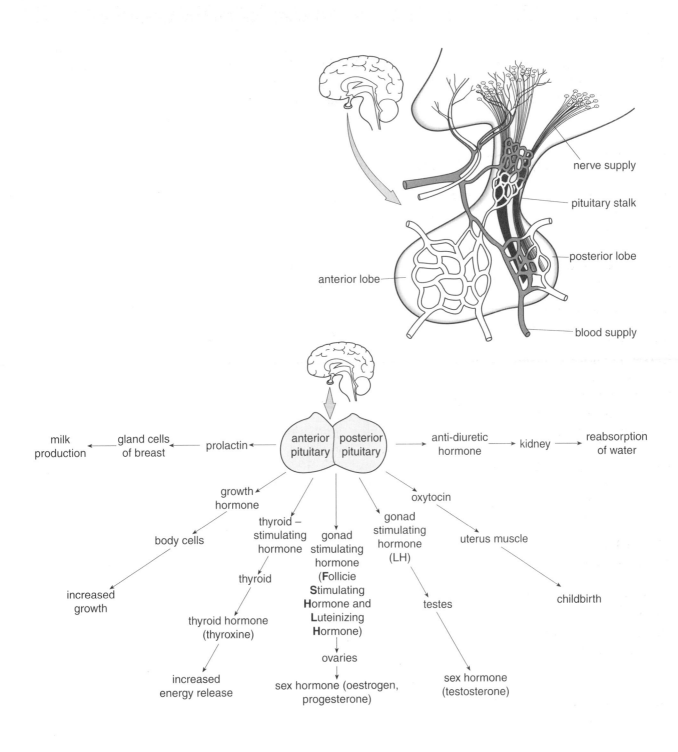

Figure 1.7.3 The pituitary gland and the effects of its hormones

production of some types of white blood cells and the structure of connective tissue. If the adrenal cortex is damaged or destroyed, a person develops **Addison's disease**. Symptoms include tiredness, nausea and weight loss.

Blood circulation is affected and skin colour changes occur. Sufferers can be helped by treatment with a corticoid hormone called **cortisone**.

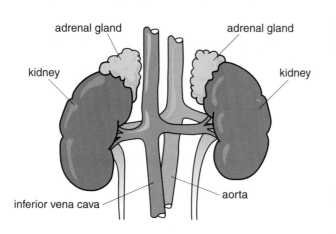

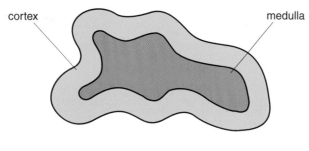

adrenal gland; cross section

The medulla secretes a hormone called **adrenalin**. It can cause sudden body changes during anger or fright. These glands are called glands of emergency because they have this effect. When a lot of this hormone is produced, the following happens:

1 You become pale, because blood vessels in your skin constrict. If you have a skin wound, you will lose less blood because of its diversion from the skin. At the same time, more blood is supplied to your muscles, brain and heart.

2 Your blood pressure goes up, because the blood vessels in your skin have constricted.

3 Your heart beats faster and its **stroke volume** increases. The stroke volume is the volume of blood pumped out of the heart at each beat.

4 Your liver releases some of its stored carbohydrate which is oxidised to release more energy during cellular respiration.

Figure 1.7.4 The adrenal glands

The pancreas

Besides producing **pancreatic enzymes** to aid digestion (see page 26), the pancreas also has special cells which produce hormones. The cells are called **islets of Langerhans** and the hormones are called **insulin** and **glucagon**. Insulin enables the liver to store glucose in the form of glycogen and speed up the oxidation of glucose in cellular respiration.

A person who does not have insulin is unable to store or oxidise carbohydrate properly. Consequently, tissues are deprived of fuel leading to a build-up of glucose in the blood. This can be dangerous for the following reason. The normal concentration of glucose in the blood is 0.1 g per 100 cm^3 of blood. If it rises above this, the kidney tubules cannot cope when they try to re-absorb the glucose back into the blood (see page 179). It is as if the tubule

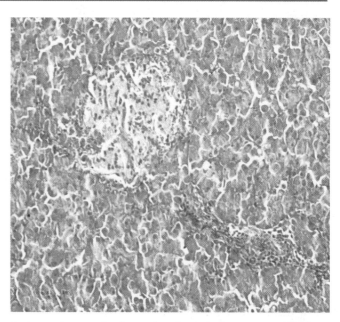

A photomicrograph of a section through the pancreas ×168

'pumps' break down. Glucose then passes right through the kidney tubules and is lost in urine.

This condition is called **diabetes mellitus** and is caused by a combination of factors. It is not just the failure of the islet cells to produce insulin. The condition can also be affected by the activity of the pituitary, thyroid, adrenals, and the liver.

There are three main types of diabetes mellitus:

1 **Juvenile**. Either insulin is not produced or the body does not react to insulin in the normal way. It is detected early in the life of the person.

2 **Maturity onset**. Confined mainly to overweight, middle-aged women and occurs when there is too little insulin to meet the demands of the body.

3 **Stress**. May occur during pregnancy or under physical or mental stress. The body fails to respond to insulin at its normal levels in the blood.

Treatment by regular injection of the correct amount of insulin normally allows people with diabetes mellitus to lead a normal life if they pay attention to carbohydrate intake in their diet.

If too much insulin is produced, the result is a condition called **hypoglycemia**. It means 'low blood sugar level' and results in tiredness. The hormone, called **glucagon**, produced in different islet cells, has the opposite effect to insulin. It encourages the change from glycogen to glucose when extra fuel is needed by the body's cells.

The ovaries and testes

Besides producing the body's sex cells, the ovaries and testes produce **sex hormones** which are used to regulate sexual cycles in the body (see page 120). The ovaries secrete **oestrogen** and **progesterone**. Testes produce **testosterone**.

Sex hormones affect the development of the **secondary sexual characteristics**. These features appear at the time of change from child to adult. They begin with the maturation of the ovaries and testes during **puberty**. Among animals, the changes may appear as the bright

feathers of most male birds, and the antlers of stags. In boys, the voice breaks and becomes deeper, and a beard and pubic hair appears. These features are accompanied by a broadening and deepening of the chest and rapid bone growth.

As a girl matures, her breasts develop and her hips become broader. Fat deposits form under the skin, pubic hair grows and menstruation begins. These physical changes cause both boys and girls to go through major mental and emotional changes besides.

The thymus

There is some evidence to suggest that the thymus might have an endocrine fuction but it is still, perhaps, the least well known gland in the body. It grows under the breastbone, just

above the heart, and continues to enlarge until the teens and then shrinks. It is certainly concerned with the body's defence mechanism because it produces **lymphocytes**.

The hypothalamus

The hypothalamus is located at the base of the brain, just above the pituitary gland. It is vital in the maintenance of the internal environment of the body. Body temperature, ion and water balance, and release of pituitary hormones are some of the functions regulated by the hypothalamus. The pituitary is directed by the hypothalamus to release the required amount of

a particular hormone. For example, if the level of water in the blood is low, it is detected by the hypothalamus which in turn directs the posterior pituitary to secrete more anti-diuretic hormone. This causes the kidney tubules to reabsorb more water in order to restore the water balance of the body (see page 11).

Body regulators

Table I A check list of the main ductless glands and their secretions

Gland	Location	Hormone	Function of hormone
thyroid	neck, below larynx	thryoxine	accelerates the rate of metabolism
parathyroids	back surface of thyroid lobes	parathormone	controls the use of calcium in the cells
pituitary (anterior lobe)	base of mid-brain	growth hormone gonadotropic ACTH lactogenic thyrotropic	regulates growth of skeleton regulates development of sex hormones in ovaries and testes stimulates secretion of hormones by adrenal cortex simulates secretion of milk by mammay glands stimulates activity of the thyroid gland
(posterior lobe)		oxytocin ADH	regulates blood pressure and stimulates smooth muscle controls water reabsorption in the kidney
adrenal cortex	above kidneys	cortin	regulates metabolism, salt, water balance, controls production of certain white cells and structure of connective tissue
adrenal medulla		adrenalin	causes constriction of blood vessels, increase in heart action and output
pancreas (islets of Langerhans)	below and behind the stomach	insulin glucagon	enables liver to store carbohydrate and regulates oxidation of glucose changes glycogen to glucose
ovaries	pelvis	oestrogen progesterone	produces female secondary sex characteristics maintains growth of uterus during pregnancy
testes	in scrotum	testosterone	produces male secondary sex characteristics

Summary

1 Ductless glands are called endocrine glands and secrete hormones directly into the bloodstream.

2 Many important hormones are produced by the thyroid, parathyroid, pituitary and adrenal glands.

3 The pancreas, ovaries and testes also secrete hormones essential for normal growth and body functions. Together, these hormones influence body metabolism, growth, mental ability, and chemical balance in the body fluids.

4 Glands are controlled by their influence on each other, by feedback and by the nervous system.

LIFE PROCESSES

Questions for review

1 How does the blood help in the work of the endocrine glands?

2 How does the thyroid gland control the rate of metabolism?

3 In what ways do the pituitary and thyroid glands affect growth?

4 In what ways are puberty and adolescence a result of glandular activity?

5 Why does sugar appear in the urine of a person with diabetes mellitus?

Applying principles and concepts

1 How do you account for the fact that the heartbeat of a footballer increases a great deal before the game as well as during the game?

2 Why is a study of the endocrine glands often carried out at the same time as a study of the nervous system?

LIFE PROCESSES

LIFE PROCESSES

The body framework

Learning Objectives

By the end of this chapter you should be able to:

- Name the systems of the body
- Name the body cavities
- Describe the structure and function of bone
- Name the kinds of joints

- Explain the theory of how muscles contract
- Describe the three types of muscle
- Explain the origin and insertion of muscles

Groups of cells

The human body is like all other animals' bodies because it is made up of cells and cell products. Groups of cells with similar functions and origins make up the different tissues of our bodies. There are two groups of such tissues.

Connective tissues

These lie between groups of nerve and muscle cells. They fill up spaces where there are no specialised cells and also form protective layers. They give strength and firmness and include muscles, tendons, ligaments, bone and blood. By definition, connective tissues have the majority of their material as a matrix in which cells are suspended.

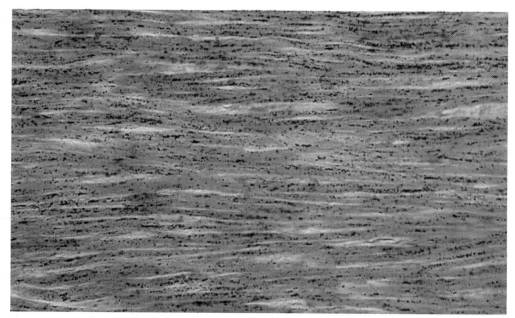

Photomicrograph of a tendon ×382

Epithelial tissues

These cover the body surfaces, inside and out. For example, one type covers the blood vessels and heart. Another type lines the stomach and intestines. Some cells of this lining are modified to secrete **mucus** and other materials that aid digestion. The skin is another epithelial tissue, as is the ciliated lining of the trachea. By definition, epithelia are layers of cells supported on a basement membrane, rather than in a matrix like connective tissue. Tissues are organised into **organs** and **organ systems**. Each organ has its own job to do and is made of several types of tissue working together for efficient functioning of the organ and, in turn, the organ system.

The body can be divided into the following ten organ systems:

1 Skeletal – bone and cartilage
2 Muscular – muscles
3 Digestive – teeth, mouth, oesophagus, stomach, intestines, liver, pancreas
4 Circulatory – heart, arteries, veins, and capillaries
5 Excretory – kidneys and bladder
6 Integumentary – skin and hair
7 Respiratory – nasal passages, trachea, bronchi, lungs
8 Nervous – brain, spinal cord, nerves, eyes and ears

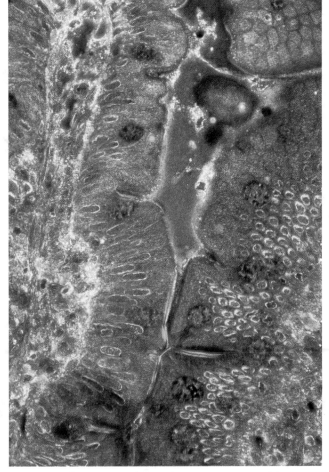

Section through columnar epithelium in a villus ×308,800

9 Endocrine – ductless glands
10 Reproductive – testes, ovaries, uterus, oviducts.

Regions of the human body

Our body is similar to that of other mammals in that we have limbs, a head, neck and trunk. The head includes the **cranial cavity** formed by the bones of the skull and encloses the brain. The head also contains many of our sense organs because they are near the brain and can send impulses to it quickly.

The **thoracic cavity** is formed by our ribs, breastbone and spine. It holds the lungs, trachea, heart and oesophagus. A dome-shaped partition called the **diaphragm** separates the thoracic cavity from the **abdominal cavity** in the lower trunk. The abdominal cavity holds most of the alimentary canal, spleen, kidneys and, in females, the ovaries.

The skeletal system

Most of our support comes from our very efficient framework called an **endoskeleton** (Figure 1.8.1). It gives maximum support with the minimum of material and allows us to move freely. However, one disadvantage to having our skeleton on the inside of the body is

that it does not offer the same protection to soft parts as a skeleton on the outside of the body would (an **exoskeleton**). Even so, many delicate parts, such as the brain, eyes and ears, are protected by the skull, and the spinal cord by the vertebrae.

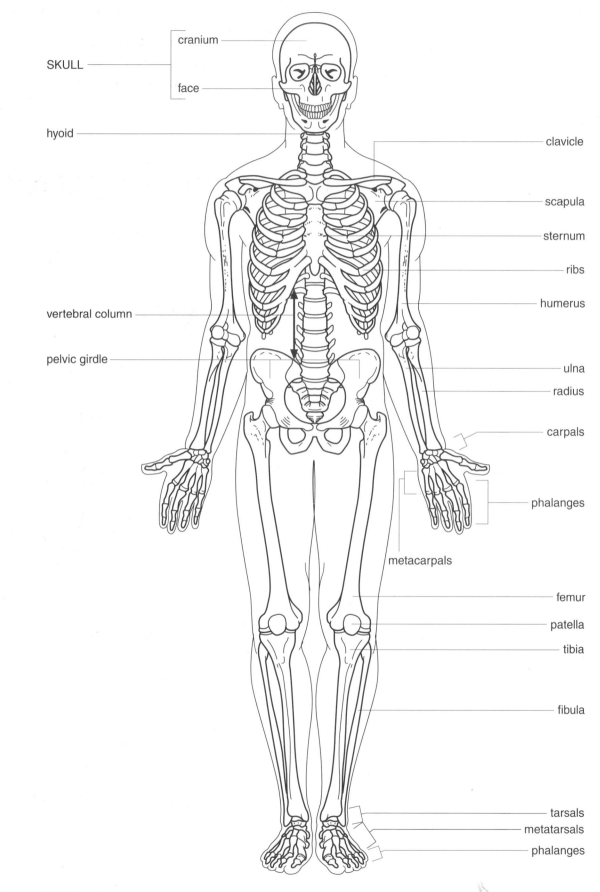

SKULL
cranium
face

hyoid

vertebral column

pelvic girdle

clavicle

scapula

sternum

ribs

humerus

ulna

radius

carpals

phalanges

metacarpals

femur

patella

tibia

fibula

tarsals

metatarsals

phalanges

Figure 1.8.1 The skeleton

Your skeleton carries out the following functions:

- Support, e.g. the vertebral column
- Provision of a surface to which muscles are attached so that movement can take place, e.g. limb bones, ribs, hip and shoulder girdles

- Protection of some delicate organs, e.g. the skull and the rib cage
- Storage of minerals, e.g. calcium is stored throughout the skeleton
- Manufacture of blood cells, e.g. in the red bone marrow of the breastbone and ankle bones.

The joints of the body

The point at which two bones meet is called a **joint**. Several types of joints occur in the human skeleton. Your elbow and knee are examples of **hinge joints** which allow movement in one plane. When the biceps muscle of the upper arm, or humerus, contracts, the lower arm can be pulled upwards only. Such joints can provide great mechanical power because there is little danger of twisting them.

Ball-and-socket joints occur at your hips and shoulders. At your hip, the ball on the end of your thigh bone, or femur, fits into a socket in the hipbone, the pelvis. This kind of joint allows movement in any direction the muscles around it allow. The humerus fits into the shoulder, the pectoral girdle, in a similar way. Tough strands of connective tissue, called **ligaments**, hold the bones in place at joints to prevent dislocation. Ligaments are elastic so can stretch to loosen a joint and let you move more easily.

Partially movable joints attach your ribs to the vertebrae in your backbone. Together with the long strands of **cartilage**, which attach some ribs to the breastbone, they allow breathing movements (see page 52).

Immovable joints occur in our skulls. A **pivot joint** connects your head to your spine and the vertebrae are connected by **gliding joints**.

All movable joints have inner surfaces lined by smooth, slippery cartilage and are lubricated by **synovial fluid**, secreted by the **synovial membrane** (Figure 1.8.3).

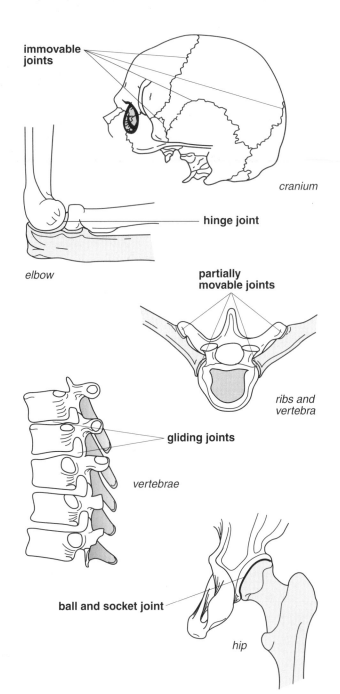

Figure 1.8.2 Different types of joints

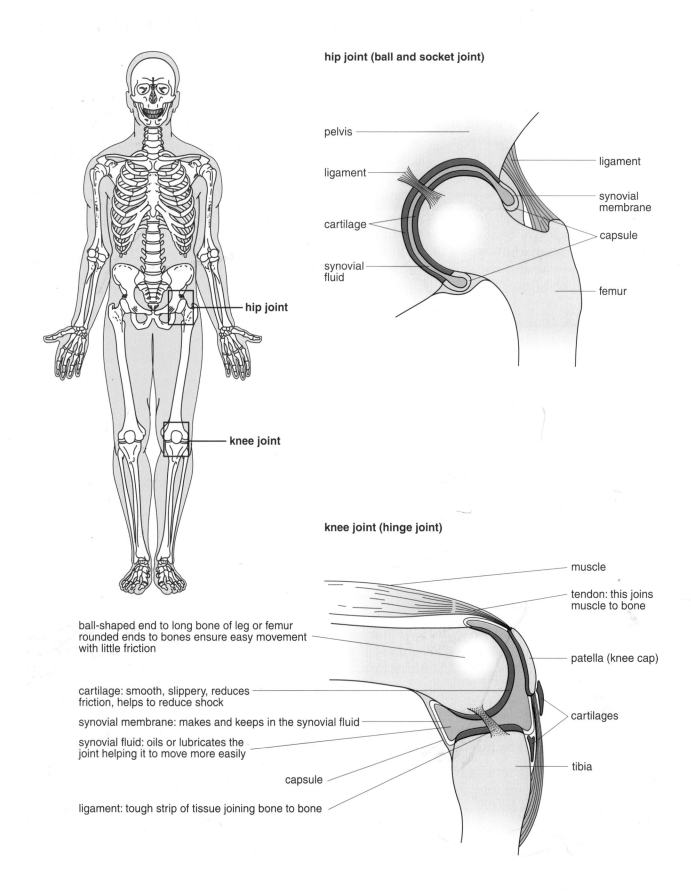

hip joint (ball and socket joint)

pelvis

ligament

cartilage

synovial fluid

hip joint

ligament

synovial membrane

capsule

femur

knee joint

knee joint (hinge joint)

muscle

tendon: this joins muscle to bone

ball-shaped end to long bone of leg or femur rounded ends to bones ensure easy movement with little friction

patella (knee cap)

cartilage: smooth, slippery, reduces friction, helps to reduce shock

synovial membrane: makes and keeps in the synovial fluid

synovial fluid: oils or lubricates the joint helping it to move more easily

cartilages

capsule

tibia

ligament: tough strip of tissue joining bone to bone

Figure 1.8.3 Synovial joints

How muscles produce movement

The bones of the skeleton need muscles to make them move. Muscle cells are specialists in movement because they are able to contract. Every movement you make is caused by the contraction of bundles of muscle cells. Besides the obvious movements you make to lift things or walk, the muscles in your digestive system and in your heart are important for **peristalsis** (see page 21) and blood circulation (see page 40). In fact, muscle tissue makes up about half of your body weight.

Muscles which move the skeleton are made of bundles of long slender cells called **fibres**. Each fibre contains many fine parallel threads called **fibrils** which run lengthwise in the fibre. There are two types of proteins called **actin** and **myosin** making up the bulk of the fibrils and are responsible for the actual process of contraction.

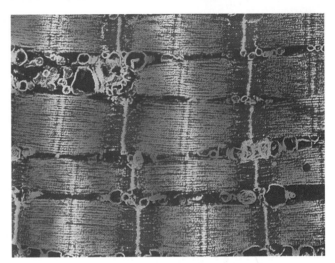

False colour micrograph of striations on muscle fibrils

Muscle fibres contract when they are supplied with energy from **adenosine triphosphate (ATP)** (see page 7), the energy currency of all cells. However, before the ATP can be used, the fibres must be activated by a **nerve impulse**. Some complex biochemistry occurs to cause the contraction, but the end product results from an attraction between actin and myosin molecules which seem to slide over each other in thousands of sections along the fibres. The whole muscle shortens because of these reactions.

Each nerve cell that carries impulses to muscle fibres branches out and forms a **motor unit**.

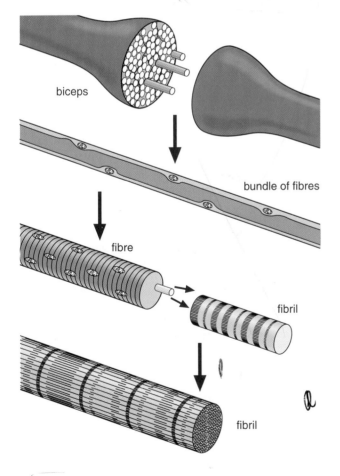

biceps

bundle of fibres

fibre

fibril

fibril

Figure 1.8.4 Muscle structure

A motor end plate in striated muscle

LIFE PROCESSES

When stimulated, each fibre contracts as tight as it can, the strength of each contraction always being the same. The strength of the movement depends on how many motor units are called into action. So the more units that contract, the greater the movement. In this way we can control very precise movements and very forceful ones. Muscle fibres can also be stimulated by heat, chemicals, pressure, and electricity.

Types of muscles

There are three types of muscle:

1 **Smooth muscle** which is made of spindle shaped cells, each containing one nucleus. It is usually in layers, e.g. the alimentary canal and arteries. All smooth muscle action is controlled by parts of the nervous system over which we have no conscious control. Smooth muscle is therefore **involuntary muscle**.

Light micrograph of striped skeletal muscle

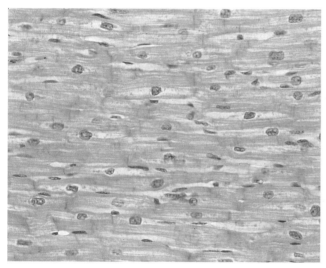

Smooth muscle ×290

2 **Skeletal muscle** which is made of bundles of fibres, each with many nuclei. Most skeletal muscles are attached to bones by inelastic tendons. As these muscles can be controlled by conscious effort, they are called **voluntary**.

In order to cause movement, muscles must attach to bones at two points and contract across joints. The attachment to the stationary bone is the **point of origin**, whereas the attachment to the movable bone is the **point of insertion**. A simple example of this is shown in the movement of the forearm.

Figure 1.8.5 illustrates **antagonistic** action of pairs of muscles. When the biceps (**flexor**)

contracts, the triceps (**extensor**) relaxes and the forearm is raised. The opposite happens when the forearm is straightened. The same action occurs between flexors and extensors in the leg.

Even when you are at rest, flexor and extensor muscles are slightly contracted. This is called **muscle tone** and the more you use your muscles, the larger they become and the more tone they have. Good posture is only possible if

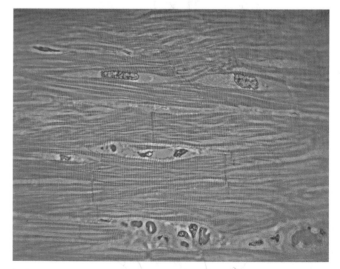

Light micrograph of cardiac muscle ×1856

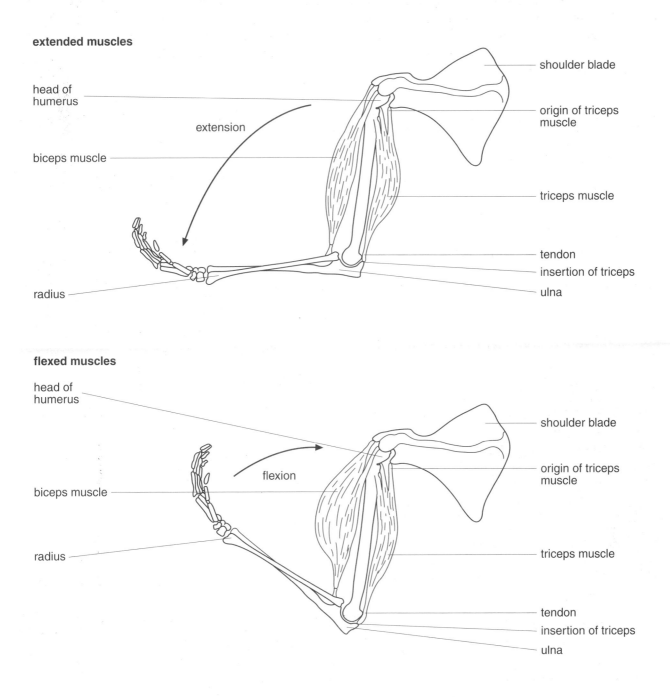

extended muscles

head of humerus

extension

biceps muscle

radius

shoulder blade

origin of triceps muscle

triceps muscle

tendon

insertion of triceps

ulna

flexed muscles

head of humerus

flexion

biceps muscle

radius

shoulder blade

origin of triceps muscle

triceps muscle

tendon

insertion of triceps

ulna

Figure 1.8.5 Skeleton and muscles in the arm

your muscle tone is good and your muscles have not become weak and flabby through under-use.

3 **Cardiac muscle** is a special type of involuntary muscle found in the heart (see page 41). The cells branch to form a network.

When cardiac fibres contract, the chambers of the heart are squeezed resulting in blood being forced out through arteries.

LIFE PROCESSES

How strong is your arm?

Procedure

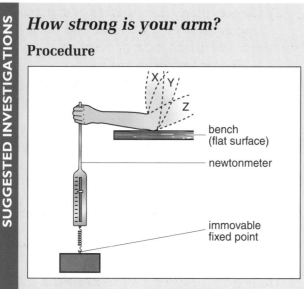

bench (flat surface)

newtonmeter

immovable fixed point

Figure 1.8.6 How to set up the experiment

Place your arm on a flat surface as shown in the diagram and get someone to hold your upper arm in position X. Fix a newtonmeter to a point which cannot move, below the bench as shown. Without moving the upper arm, pull on the newtonmeter and record the maximum force you can exert. After resting the arm, repeat the procedure with the upperarm held in positions Y and Z.

Which is the best position to hold your arm when you lift things? Explain your answer.

Questions for review

1 What are the three body cavities?

2 What are the functions of the bones? Give an example of a bone serving each purpose.

3 Describe the tissues found in a joint and state their functions.

4 Describe the types of joints found in humans.

5 Name the three types of muscle tissue and state where they are found.

6 Explain antagonistic muscle action.

Applying principles and concepts

1 Explain the importance of a highly developed nervous system in an organism with an internal skeleton.

2 Suggest why the ankle bones of birds are fused together and why a common sports injury in humans is a sprained or dislocated ankle.

Summary

1 Our bodies are very similar to those of other complex animals.

2 Our organs are made of tissues, the tissues are made of cells which vary depending on their function.

3 Unlike simple animals, we have bony skeletons.

4 Our skeletons have several functions including support, movement, protection, blood cell production, and mineral storage.

5 Muscles produce movement of the skeleton by being attached to bones and acting across joints.

6 Smooth muscles form layers in the walls of organs like the stomach, intestines and arteries.

7 Special cardiac muscle is used in the heart when it pumps blood.

Sample examination questions

1 The diagram below is of a section through a canine tooth in the jaw.

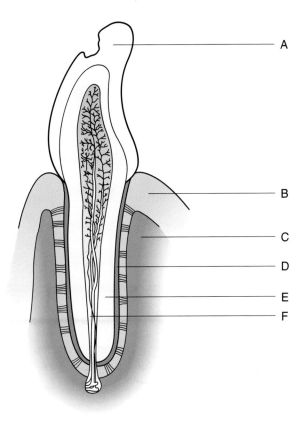

a) Choose letters from the diagram to fit the description below. Each letter may be used once, more than once, or not at all.
  i) The hardest part of the tooth.
  ii) The part of the tooth which contains blood vessels.
  iii) The part which holds the tooth in the jaw.
  iv) The part of the tooth on which plaque develops.

b) The tooth shown is decayed. Explain why the person with this tooth would feel no pain.

WJEC, 1993

2 The diagram below represents part of the digestive system of a human.

a) i) On the diagram identify the connection of the pancreas to the alimentary canal.
  ii) Select the correct words from the list below to fill in the boxes in the diagram.

**hepatic artery   hepatic vein
glycogen   sugar   urine   urea
ileum   pancreas**

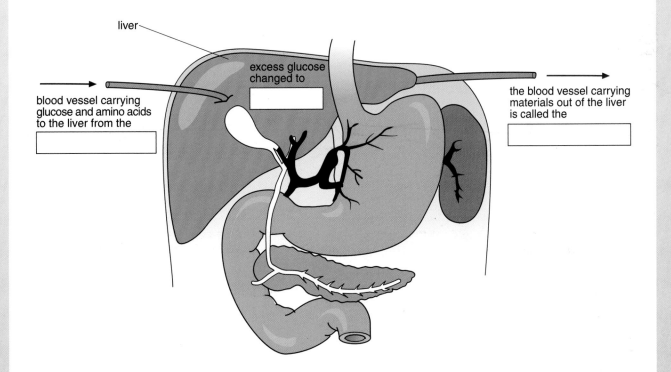

liver

excess glucose changed to

blood vessel carrying glucose and amino acids to the liver from the

the blood vessel carrying materials out of the liver is called the

b) Copy and complete the table below to show the chemical elements present in the foods listed. The first one has been done for you.

Class of food	Chemical elements present
carbohydrate	carbon, hydrogen, oxygen
protein	
fat	

c) Identify the correct answer below. Digestive enzymes are essential because they
  i)   are protein
  ii) are unchanged after a reaction
  iii) can help break down large molecules
  iv) are denatured at high temperatures.

d) If the normal body temperature is 37°C, which of the graphs below is correct for enzyme activity in the human body?

WJEC, 1994

3 There is evidence to suggest that a high cholesterol level in the blood is linked to diet. Graph A shows the relationship between death rate and cholesterol levels in the blood of people in the USA.

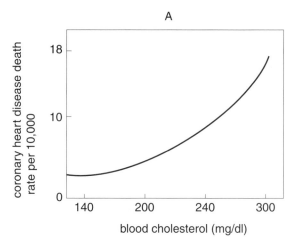

A

a) What can you conclude from Graph A?

b) Graph B shows the amount of cholesterol in the blood of Americans aged between 20 years and 74 years.

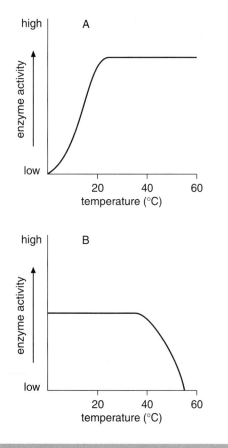

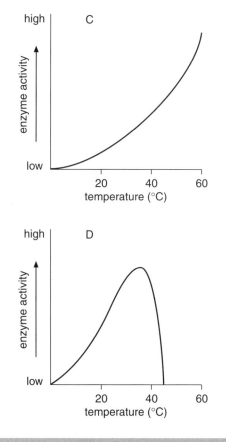

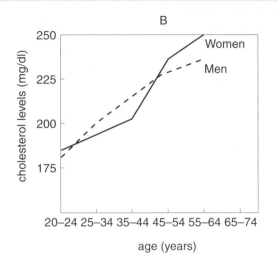

B

cholesterol levels (mg/dl)

age (years)

Use Graph B and the given information to answer the question which follows. Substances called Low Density Lipoproteins (LDLs) carry cholesterol in the blood and release it on the inside of the arteries. High Density Lipoproteins (HDLs) take cholesterol out of the blood and away from the walls of the arteries.

A female sex hormone (oestrogen) reduces levels of LDLs in the blood and increases HDLs. On average, females stop producing oestrogen between the ages of 50 and 60.

Explain, fully, the health risk from heart disease in these women.

WJEC, 1994

4 The diagram below shows a 'model gut' which can be used to investigate why starch must be digested.

The dialysis tubing has different properties from the lining of the small intestine.

a) i) What is represented by the contents of the tubing?
   ii) What is represented by the distilled water?
   iii) At what temperature should the water bath be set?

b) After 20 minutes the contents of the boiling tubes and of the dialysis tubes were tested for the presence of starch and reducing sugar.
   i) What chemical would be used to test for the presence of starch?
   ii) What colour would indicate the presence of starch?
   iii) How would you carry out a test for the presence of reducing sugar?

c) The results of these tests are shown in the table below.

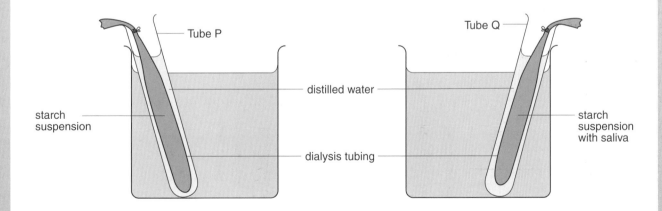

Tube P

starch suspension

distilled water

dialysis tubing

Tube Q

starch suspension with saliva

	Tube P		Tube Q	
	water	dialysis tube	water	dialysis tube
starch	absent	present	absent	present
sugar	absent	absent	present	present

Use these results to explain why starch in food must be digested.

NEAB, 1994

5

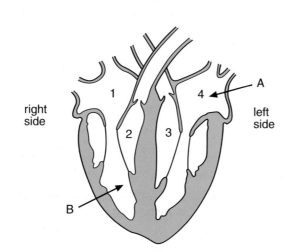

right side

left side

a) i) Copy the diagram of a dissected heart and label parts A and B.
   ii) Draw arrows at points 1, 2, 3 and 4 to show the direction of flow of blood on your diagram of a dissected heart.

b) The table shows the pressure readings in parts of the heart while it is pumping blood.

	Pressure (kPa) in left ventricle contracting	Pressure (kPa) in left ventricle relaxing
left ventricle	16.0	0.7
left atrium	0.4	1.1
aorta	12.0	10.7

Use the information in the table to explain what happens in terms of pressure changes and blood flow when the:
i) Valve between the left atrium and the left ventricle opens.
ii) Valves guarding the base of the aorta open.

WJEC, 1995

6 The diagram below shows the breathing organs inside the human chest.

a) On the diagram, name the parts labelled A and B.

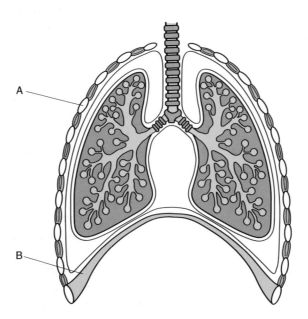

b) i) What makes part A move?
   ii) Which way does part B move when we breathe in?
   iii) Explain why air moves into the lungs when parts A and B move.

c) Explain why the rate of breathing increases during exercise.

NEAB sample material, 1995

7 a) The diagram shows nephrons in part of the human kidney.

b) The table shows the percentage of some chemicals in blood plasma, glomerular filtrate and urine.

Use the information in the table to answer the following questions.

i) What substances are passed from the blood plasma to form the glomerular filtrate?
ii) Describe the difference in percentages of chloride ions and urea, between the blood plasma and urine.

Substance	Percentage in blood plasma	Percentage in glomerular filtrate	Percentage in urine
water	90.00	90.00	95.00
protein	7.00	0.00	0.00
glucose	0.10	0.10	0.00
chloride ions	0.40	0.40	0.60
urea	0.03	0.03	2.00

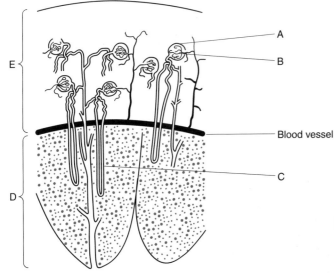

iii) Explain why the glucose percentage in urine is zero.
iv) Suggest one condition in which urine contains glucose.
v) An increase in the protein intake in the diet of an adult causes a corresponding increase in the amount of urea in urine. A similar increase in the protein intake of the diet of a teenager does not cause a similar increase in the amount of urea in urine. Explain this difference.

NICCEA, 1995

i) Name the parts A, B, C, D and E.
ii) Copy the diagram and label the urine collecting duct and mark, with arrows, the direction of flow.
iii) Explain the process of ultrafiltration.

8 The graph overleaf shows the volume of urine released by a man over a period of several hours. The urine was collected every half hour and its volume measured. At point X the man drank a litre of cold water.

a) i) What was the rate of output of urine per half hour before the man drank the water?
   ii) Describe precisely the effect on urine production, over the $3\frac{1}{2}$ hours after drinking the water.
   iii) How long did it take his rate of urine production to return to normal after having the drink?

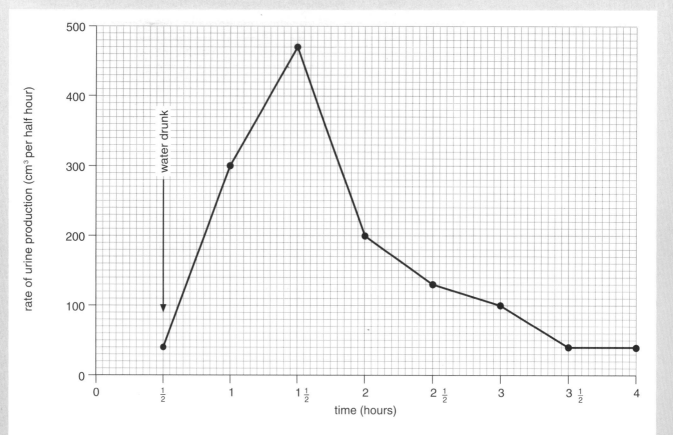

iv) What was the largest amount of urine formed in half an hour?

b) This investigation was carried out in cold conditions during winter.
   i) If the investigation was repeated during a hot summer's day, give two differences which you would expect the graph to show compared with the above.
   ii) Briefly explain the reasons for these differences.

WJEC, 1995

9 The diagram opposite (top) shows the pathway taken by nerve impulses which help the very quick change in size of the pupil to be made.

a) Name this type of pathway.

b) What would be the stimulus for the reaction to take place?

c) Copy the diagram and label the receptor, co-ordinator, and the effector.

d) In an investigation into nervous stimulation of muscles, a scientist found that when a muscle was stimulated it first contracted, and then returned to its normal length. She measured the way it contracted at three different temperatures and the results are shown on the graphs opposite.

State three effects of temperature on the muscle contraction and relaxation.

WJEC, 1995

10 The diagram on page 112 (top) shows a section across the spinal cord.

If C is injured then part of the person's body is unable to move (paralysis) and there is also loss of feeling. If A is injured there is a loss only of feeling, movement is still possible. Injury to B causes paralysis but there is still a sense of feeling.

Suggest an explanation for these effects.

SEG, 1993

11 The liver is an organ of homeostasis.

a) What is meant by the term, homeostasis?

b) Name one substance found in the blood which is regulated by the liver.

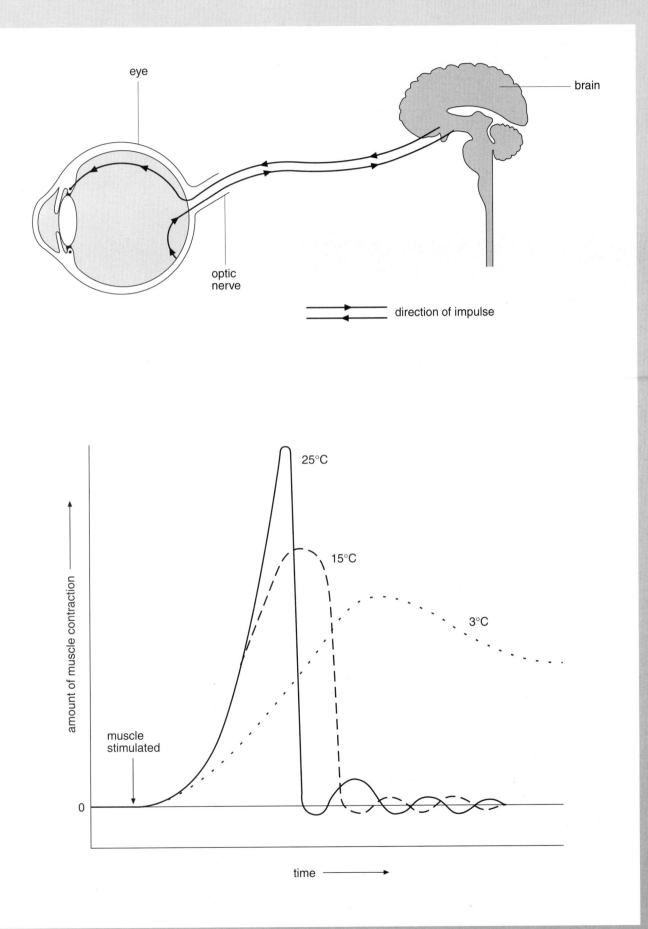

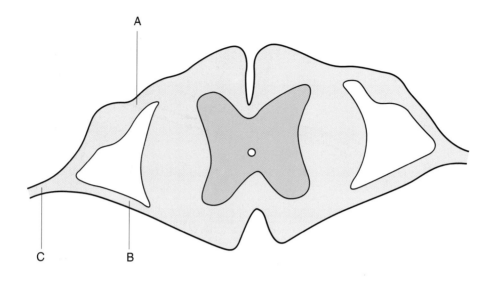

c) Name two hormones, produced by the pancreas, involved in homeostasis.

d) State the function of each hormone.

e) The normal concentration of glucose in the blood is 0.1 g per 100 cm³. Suggest one reason why it is dangerous for the concentration of glucose in the blood to drop below this level.

f) The flow diagram below demonstrates the principle of negative feedback to maintain the balance of materials in the body.

Copy and complete the flow diagram opposite to show how the production of a hormone by the pancreas demonstrates negative feedback.

WJEC, 1994

12 a) The diagram opposite shows an elbow joint.
   i) Copy the diagram and label parts A-D.
   ii) State the function of part B.
   iii) On the diagram, draw and label the cartilage in the correct positions.

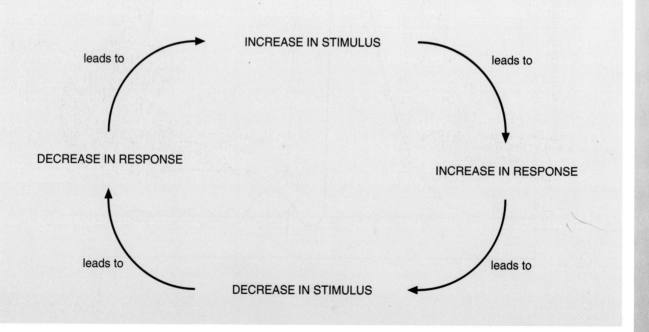

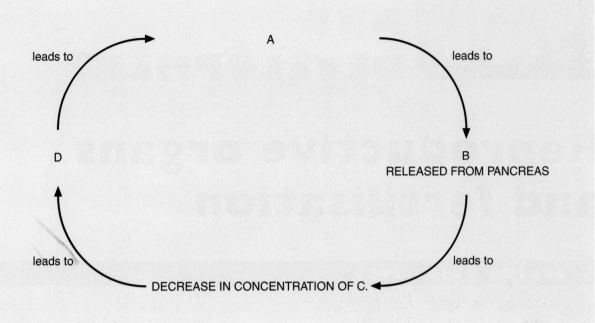

leads to

A

leads to

D

B
RELEASED FROM PANCREAS

leads to

leads to

DECREASE IN CONCENTRATION OF C.

iv) Describe two features of the joint which help it to move smoothly.

v) Explain how the action of pairs of muscles causes the bones to move at the joint.

b) When a joint is damaged, medical technology can be applied by fitting an artificial joint as shown in the diagram below.

Give three visible differences between the normal joint and the artificial metal joint.
WJEC, 1994

LIFE PROCESSES

NORMAL JOINT

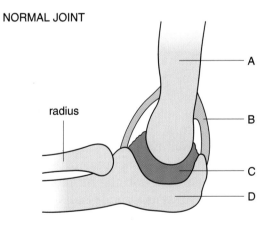

radius

A

B

C

D

ARTIFICIAL JOINT

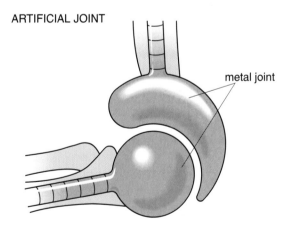

metal joint

REPRODUCTION AND DEVELOPMENT

Reproductive organs and fertilisation

Learning Objectives

By the end of this chapter you should be able to:

- Explain the significance of sexual reproduction
- Describe the male and female reproductive organs
- Understand the functions of the reproductive organs
- Describe the sperm cell

- Explain the stages of the menstrual cycle
- Understand the significance of fertilisation
- Understand the major methods of contraception

The significance of sexual reproduction

Basically, sexual reproduction is the fusion of a **male sex cell (a sperm)** with a **female sex cell (an ovum)**. These special cells are called **gametes** and the product of their fusion is a **zygote** which grows into an **embryo** and then into a mature form which resembles its parents. The process of fusion of sex cells is called **fertilisation**. Organisms which reproduce asexually – in other words, without fusion of sex cells – cannot show much variation (see page 156). However, the mixing of genetic material (genes) from two different parents can produce almost endless variation.

Sex organs, called **gonads**, produce gametes during a special kind of cell division called **meiosis** (see page 135) which results in mixing of genes.

The need for specialised structures

As with all mammals, the method of producing young in humans occurs by sexual reproduction. Normal body cells have 23 pairs of chromosomes (see page 137), and when fertilisation occurs, the zygote must also possess the correct number of chromosomes, i.e. 23 pairs. In order to acheive this, eggs and sperms must contain half the number of chromosomes present in body cells. The process of meiosis makes halving of the number of chromosomes possible (see page 137). The body requires a particular place where cells can divide during meiosis. These places are the **primary sex organs**.

Once the gametes have been produced by males and females, there must be a method for ensuring that sperms can travel safely and meet the egg so that fertilisation can occur. Also, the zygote produced inside the female's body must be protected and nourished until it develops sufficiently for birth to take place. The physical characteristics of both male and female must allow for this.

Whereas primary sex organs are special regions where meiosis takes place – the gonads – **secondary sex organs** ensure safe travel of sperm from male to female and also provide a place for the development of the new individual.

Primary sex organs

In the male, the primary sex organs are the **testes** (Figure 2.1.2).

The two testes develop in the abdominal cavity in the embryo. About two months before birth they descend through an opening in the pelvis into the **scrotum**, a sac of skin which hangs loosely between the legs (see Figure 2.1.2). The testes are egg-shaped and can differ in size from one person to another.

Since the scrotum is outside the body, its temperature is slightly lower than that of the rest of the body. The lower temperature is needed for the best conditions for the production of sperms. In addition to this method of temperature regulation, the testes are supported by muscle tissue which can contract or relax. In cold weather the muscle contracts, drawing the testes nearer to the warm body. The reverse happens in warm weather, so the process works like a thermostat.

The testes consist of a mass of sperm-producing tubes called **seminiferous tubules** which you can see in the photograph (below).

Meiosis occurs in these tubules resulting in the production of many millions of sperms, each consisting of a nucleus surrounded by a little cytoplasm which extends to form a long tail. While inside the seminiferous tubules, the sperms cannot yet move using their tails (Figure 2.1.1).

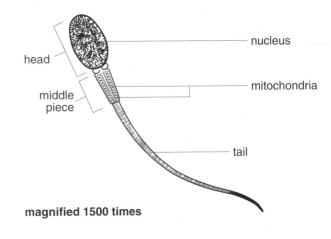

magnified 1500 times

Figure 2.1.1 A sperm

A sexually mature male can produce sperms continuously. This is possible because the sperm-producing cells in the seminiferous tubules can continue to undergo meiosis throughout a man's life from puberty (at about 12 years of age) to well over 70 years of age.

Between the seminiferous tubules are the **interstitial cells** which can produce the hormone, **testosterone** (see page 119). This hormone is responsible for the production of sperm and for the development of **secondary sexual characters** in the male.

In the female, the primary sex organs are the **ovaries** (Figure 2.1.3).

They develop in the abdominal cavity as a pair of gonads like the testes but, as development proceeds, the ovaries remain inside the

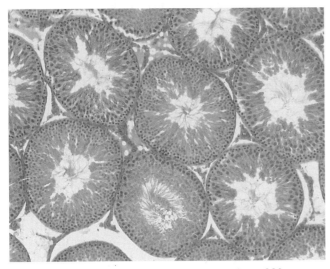

A transverse section of the seminiferous tubules ×290

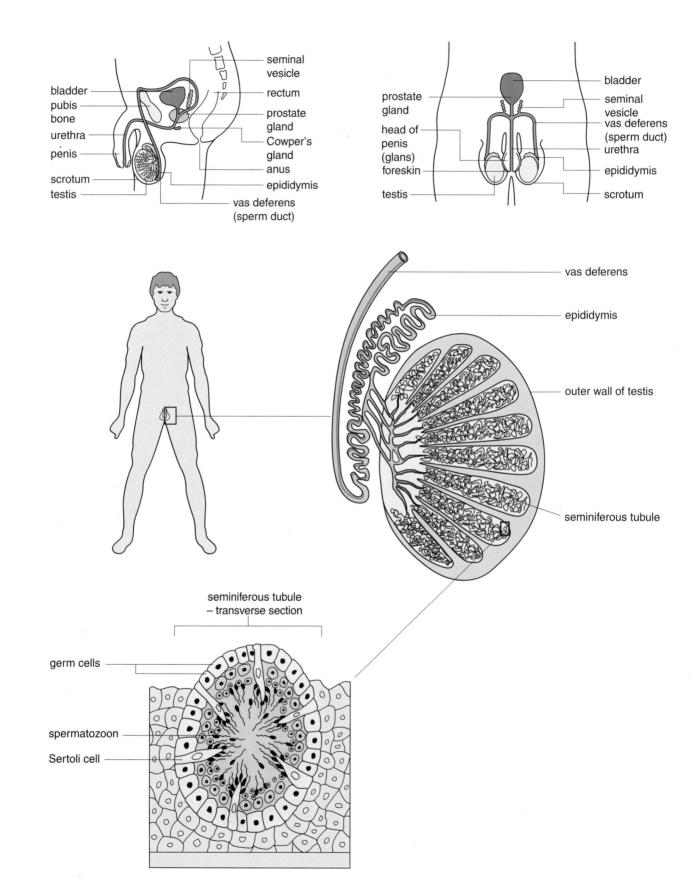

Figure 2.1.2 Male sex organs

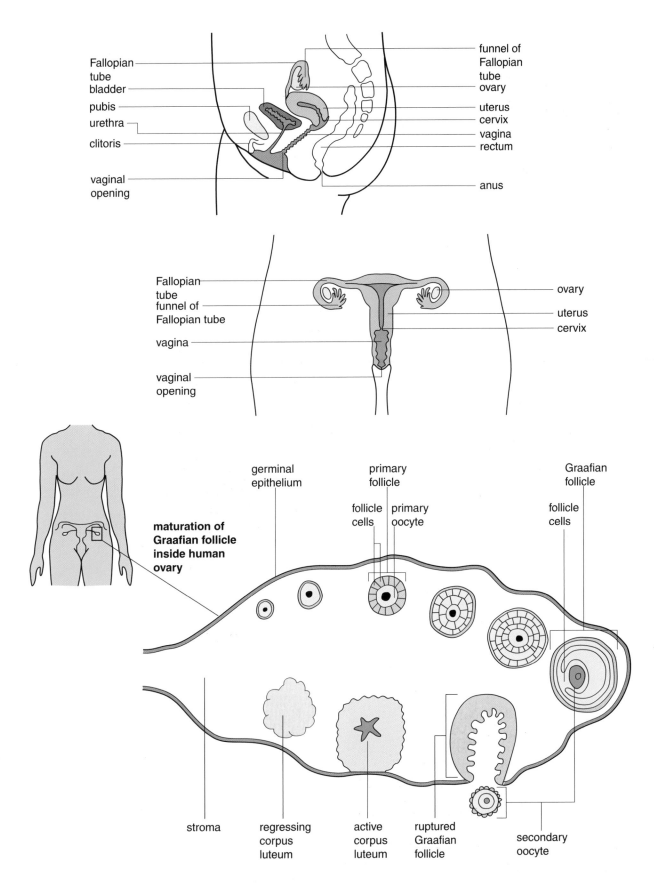

Figure 2.1.3 Female sex organs

abdomen while the testes descend into the scrotum. They are roughly the same size and shape as plums and contain connective tissue, blood vessels and potential eggs. At birth the two ovaries of the female infant contain nearly 500 000 cells which can develop into mature eggs. Of these cells, only about 400 will actually reach maturity and become capable of fertilisation.

From the age of about 10 years (the precise time of life varies greatly between one girl and another), some of the potential eggs begin to grow in fluid-filled sacs called **Graafian follicles**. If we look at the ovary of a woman, we can see follicles at different stages of development (Figure 2.1.3). Usually only one follicle becomes fully mature each month and releases an egg during **ovulation**. Once this happens, the remaining follicles and their eggs disintegrate and a new set of follicles then begin to grow, so that another mature egg can be released a month later. The cycle, called the **menstrual cycle**, is described on page 120.

Secondary sex organs

From the male's seminiferous tubules, millions of sperms pass into larger tubes called the **epididymis** in which they remain for as long as six weeks while the final stages of development are completed. When fully mature, the sperms leave the epididymis by using the whip-like action of their tails to swim through a fluid. Eventually the epididymis leads to the main tubes for the transport of sperms. These are the **vas deferens** (Figure 2.1.2) which lead to the body cavity and the top of the **prostate gland**. This gland is walnut-sized, firm, muscular, and situated immediately below the bladder. Its function is to produce **seminal fluid** in which sperms swim.

Inside the prostate gland, each vas deferens is joined by a duct from the **seminal vesicles**. These vesicles secrete a fluid which activates and provides nourishment for the sperms.

Where the duct from the seminal vesicle joins the vas deferens, it forms the **ejaculatory duct** (Figure 2.1.2). This duct passes through the prostate gland and both ejaculatory ducts open into the **urethra** (see page 63) which carries both urine from the bladder, and sperm from the testes to the outside through the **penis**.

On each side of the urethra, just below the prostate gland, are two pea-sized **Cowper's glands**. During sexual excitement these produce a few drops of fluid which lubricate the male's urethra for the passage of sperm. The final mixture of sperm and seminal fluid which passes into the urethra is called **semen**.

In the female, when an egg bursts from a follicle in one of the ovaries, it is transferred to one of the two **Fallopian tubes** (sometimes called **oviducts**) (Figure 2.1.3). They are each about 10 cm long, extending from the region of the ovaries to the upper part of the uterus. An egg is usually transferred to the Fallopian tube nearest to its corresponding ovary. It is drawn into the Fallopian tube by the wafting action of cilia which can be seen in the photo (below).

The egg, unlike the sperm, does not have its own method of locomotion and so relies on these cilia to reach the site of fertilisation in the Fallopian tube. Once the egg is in the Fallopian tube, **peristalsis** (see page 21) helps its progress together with the action of cilia.

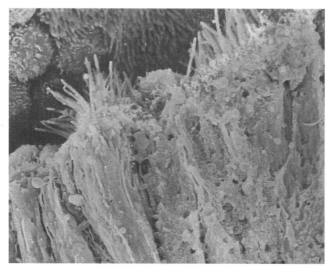

The lining of a Fallopian tube

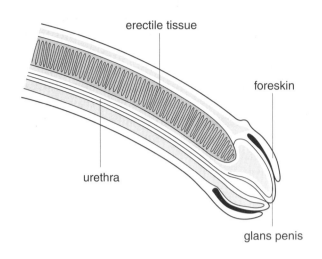

erectile tissue

foreskin

urethra

glans penis

Figure 2.1.4 An erect penis

Both sexes have the secondary sex organs which enable the transfer of sperm from a man to a woman. In the case of the male, the organ is the penis, while the female has a **vagina**.

The shaft of the penis contains the urethra and a number of layers of erectile tissue. At the end of the penis is an area which is extremely sensitive to touch, called the **glans penis**. This is covered by a loose flap of unattached **foreskin** (see Figure 2.1.4). In the operation of circumcision, the foreskin is surgically removed.

During sexual excitement changes in blood flow through the small arteries and veins of the penis cause the spaces within the erectile tissue to fill

with blood under pressure. The blood thus trapped causes the penis to enlarge and stiffen so that it can be placed into the vagina of the female.

The vagina is a muscular tube into which the penis can fit during **sexual intercourse (copulation)**. It leads from the base of the **uterus (cervix)** to an opening between the legs. The opening of the vagina is partially closed by a membrane called the **hymen**. This is broken when sexual intercourse first takes place or when a tampon is inserted for the first time during a menstrual period (see page 120). The openings of the urethra and vagina are separate in the female and just above the opening to the urethra is the small, round, elongated **clitoris**. Like the penis, this has a shaft and a glans; the shaft contains erectile tissue, and the glans is highly sensitive. Stimulation of both the glans penis and the glans of the clitoris can result in the sexual enjoyment of **orgasm**.

Once the penis is inside the vagina, further stimulation causes **ejaculation**. In this stage of sperm transfer, muscular contractions around the urethra in the erect penis cause orgasm in the male when semen is forced out of the penis and into the vagina. During the process of intercourse the female may equally experience this peak of sexual enjoyment.

Both penis and clitoris develop in the same way during early stages in the growth of the foetus. The effects of hormones determine the fate of the organs in males and females.

Control mechanisms in gamete production

The production of gametes in both male and female occurs under the control of the **endocrine system** (see page 88). In the male, the **pituitary gland** at the base of the brain secretes **luteinising hormone (LH)** which travels to the testes in the bloodstream where it causes the interstitial cells to produce the hormone **testosterone**. This is responsible for the proper maturation of sperm. Very little is secreted before puberty, which takes place between the ages of 10 and 15 years. Once the secretion begins, it continues throughout a man's life.

The female system is far more complex and operates on a monthly cycle. In this cycle one egg is released from one of the two ovaries approximately every 28 days, although this time differs for different women.

The cycle begins when the pituitary gland secretes **follicle-stimulating hormone (FSH)**. It causes a number of ovarian follicles to secrete another hormone called **oestrogen**. The effect of oestrogen is to stimulate the repair and thickening of the lining of the

REPRODUCTION AND DEVELOPMENT

uterus (**endometrium**) after the previous menstrual period. When the concentration of oestrogen builds up in the blood it affects the pituitary by causing the release of **luteinising hormone**. This reaches the ovaries in the bloodstream where it causes the mature follicle to rupture and release an egg in the process of ovulation. This occurs about 14 days after the beginning of the previous menstrual period.

Once the egg has burst from the follicle, the now empty follicle collapses and the luteinising hormone causes further changes in its structure. At the same time, its colour changes from white to yellow. The follicle has now changed into a **corpus luteum (yellow body)**. The cells of the corpus luteum continue to produce oestrogen and, also, another hormone called **progesterone**, which helps to stimulate the build-up of the lining of the uterus.

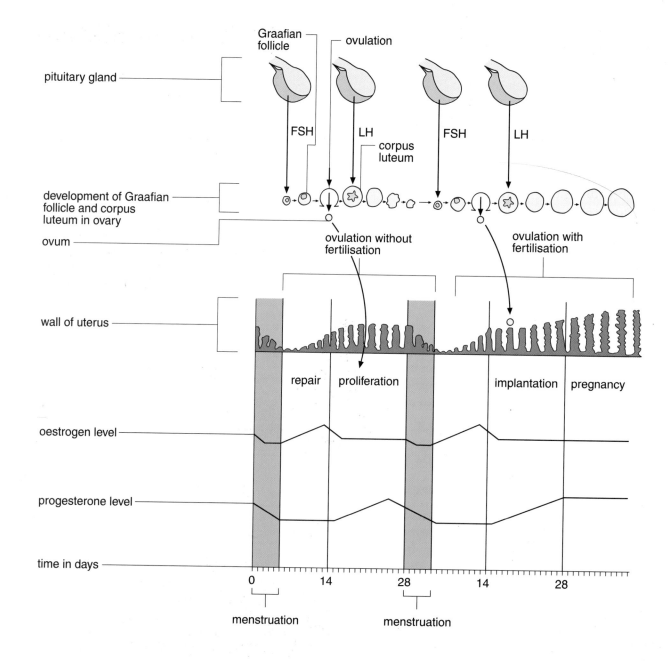

Figure 2.1.5 The menstrual cycle

Meanwhile, the egg begins its journey towards the uterus in a Fallopian tube. If it is not fertilised, the corpus luteum disintegrates. This occurs about 22 days after the beginning of the previous menstrual period and causes a decrease in levels of oestrogen and progesterone in the blood. By day 28, there is not enough of these two hormones to keep the lining of the uterus growing, so it breaks down. The resulting menstrual blood flow usually lasts for about five days, at the end of which FSH is again secreted from the pituitary.

If the egg is fertilised, the corpus luteum continues to grow and produce higher concentrations of progesterone in the blood. In this case the lining of the uterus becomes progressively thicker and more spongy, ready to receive the zygote at **implantation**. The corpus luteum continues to produce progesterone long after the fertilised egg is implanted, and the uterine wall continues to thicken.

The process of fertilisation

When ejaculation occurs during sexual intercourse, between 200 and 500 million sperm are deposited in the vagina. The acidity of the vagina is not favourable to the survival of sperm, so many die there. The remainder move towards the cervix and only about 10 000 may reach it. Further death of sperm means that only a few thousand enter the **uterus**. A few hundred may be all that finally complete the journey to the Fallopian tubes, the sites of fertilisation.

The egg is thought to be capable of fertilisation up to 48 hours after its release from the ovary and the sperm deposited in the vagina have a maximum life expectancy of 72 hours.

Fertilisation, once thought to be a simple matter of a sperm fusing with an egg, is in fact a complex series of events. When a sperm head penetrates the jelly-like coat and double membrane of the egg, further sperm are prevented from fertilising the same egg. This is caused by the immediate thickening of the outer membrane.

The 23 chromosomes of the sperm join with the 23 chromosomes of the egg to produce the normal 46. Two of these chromosomes are the sex chromosomes (see page 143). The egg always contains the X chromosome but a sperm will contain either an X or a Y. If an X fuses with an X, the child will be female; if an X fuses with a Y, the child will be male.

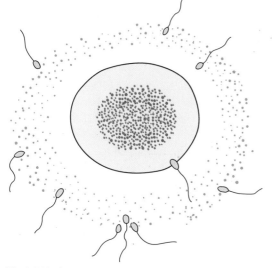

magnified 190 times

Figure 2.1.6 Fertilisation

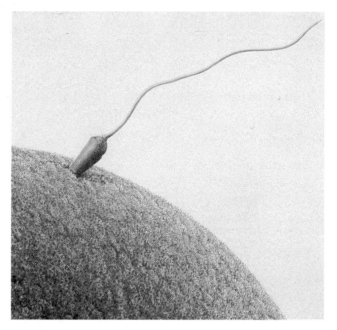

Sperm fertilising an egg

Twins

There are two kinds of twins, those that arise from a single egg and those that result from two separate eggs.

Sometimes, instead of the usual single egg being released from an ovary at ovulation, two or more are released. These may be fertilised by different sperm and will develop in the uterus at the same time. They have no greater chance of sharing characteristics than do any pair of children with the same parents. Twins that result from different eggs are called **fraternal** or **non-identical**.

Identical twins develop from a single egg after fertilisation. The zygote splits into two and each part then develops into separate individuals which share the same placenta. Each twin has identical genetic information and therefore common genes.

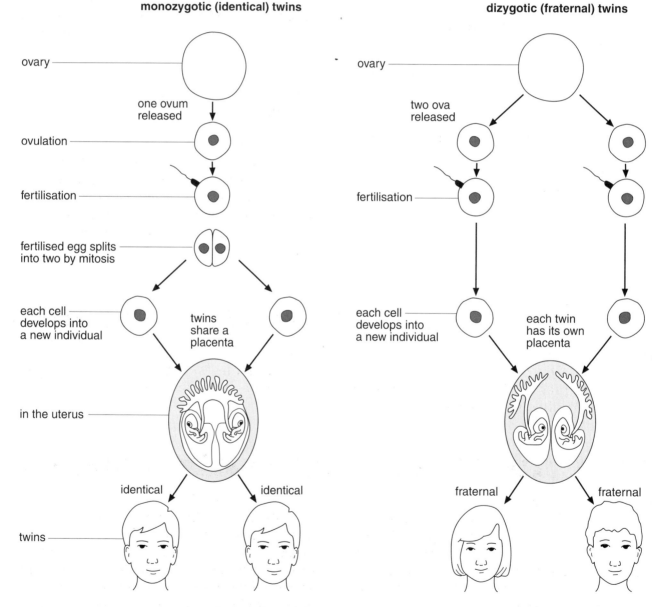

Figure 2.1.7 Identical and fraternal twins

Prevention of fertilisation

The solution to the problems caused by overpopulation lie largely in control of the birth rate by **contraception** (Figure 2.1.8). This is the prevention of fertilisation or the development of the zygote by a variety of temporary means. Birth control may also involve the **termination** of pregnancy.

Techniques of contraception may be used not only to control population growth as a whole, but also to plan individual families so that children are born into situations where they can receive loving care and attention.

Contraceptive methods can be considered as four types:

- A barrier preventing the union of sperm and egg
- A device preventing the implantation and development of the zygote
- A substance preventing the production of sperm or eggs
- A schedule of sexual intercourse that prevents fertilisation.

Barriers

One of the most commonly used devices in the developed nations is the **condom**. This is a sheath made of strong, thin rubber which is worn over the erect penis. At the open end there is a thin rubber ring which fits firmly around the penis, while at the closed end is a small pouch which provides a space for ejaculated semen. There is always a chance that the condom will slip off or split during sexual intercourse so, to increase safety, sperm-killing creams may be applied around the rim of the condom. The **femidom** is the female equivalent of a condom and can be inserted into the vagina as a barrier.

Also, women can use a **diaphragm** or **cap**. This is a thin, rubber, dome-shaped cup with a thin, metal ring around its margin. The diaphragm is designed to fit over the cervix, and the woman places it in position before intercourse. However, it is difficult to ensure that there are no gaps around the edge of the diaphragm. To prevent sperm from swimming around the edge, it is necessary to place sperm-killing cream around the margin of the diaphragm before insertion.

Prevention of implantation

If fertilisation is allowed to occur, the implantation of the zygote in the wall of the uterus can be prevented with an **intrauterine device** or **IUD** (also called a **loop** or **coil**). It is a small plastic and/or metal structure placed inside the uterus. A doctor inserts it and it can remain in the uterus for a number of years.

Prevention of gamete production

A widely used method of contraception in Britain is '**the Pill**'. This method prevents ovulation with the use of sex hormones. As the concentration of oestrogen produced by follicles in the ovaries rises, this inhibits the production of FSH from the pituitary. No development of new follicles can take place, and no development of new eggs can occur. Later in the cycle, progesterone has the same effect on the pituitary.

The contraceptive pill contains progesterone and oestrogen in different proportions. It works on the principle that progesterone and oestrogen prevent the production of FSH and so prevent the formation of new eggs. Because the concentrations of oestrogen and progesterone are kept at high levels, it is said that the Pill produces a false pregnancy.

spermicidal cream/jelly

pills

spermicidal films

condoms

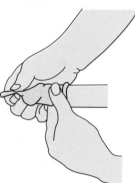

cap

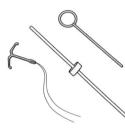

examples of IUDs

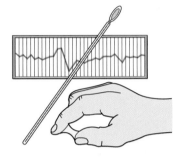

natural methods include regular taking of temperature and checking vaginal mucus

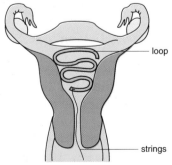

sponges

sheath slipped onto erect penis

diaphragm being inserted into vagina

cervix
vagina

diaphragm in place

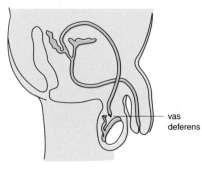

male sterilisation

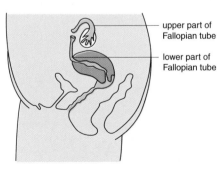

female sterilisation

upper part of Fallopian tube

lower part of Fallopian tube

intra-uterine device

loop

strings

vas deferens

Figure 2.1.8 How some methods of contraception are used

Pills are usually taken for the full 28 days each month. Even a single forgotten pill may allow pregnancy. However, it is the most reliable method of contraception but, despite its safety, the Pill is not used by some women because of the fairly drastic side-effects it can have: headaches, irritability, change of personality, and soreness of breasts are some of these. In some cases, thromboses (blood clots) have been attributed to certain contraceptive pills. A 'Pill' for men is still in a research and development phase and relies on preventing the correct development of sperm.

A schedule of sexual intercourse to avoid fertilisation

The system is called the **rhythm method** and involves abstaining from sexual intercourse during the days when the egg is available in the Fallopian tube. Intercourse is avoided for about three days on either side of the day of ovulation.

A clue to the time of ovulation is given by a woman's body temperature. Usually the temperature is slightly lower during the first 13 days of the cycle. At the time of ovulation the temperature drops slightly and is followed by a sharp rise of about 0.7°C. Another clue is in observing the quality of mucus secretions from the cervix at the time of ovulation. There is difficulty in allowing for variation in the menstrual cycle, so the method is not particularly reliable.

Permanent methods of birth control

Permanent methods of contraception are by cutting and tying both vas deferens in the male, or the Fallopian tubes in the female. This is known as **vasectomy** in males and **sterilisation** in females.

Summary

1 Sexual reproduction combines the genetic material from two different parents and this leads to greater variation.

2 Fertilisation takes place internally, the testes of the male producing gametes called sperms which are released in semen during intercourse.

3 The egg is very large compared to sperms and contains all the food for the early stages of the zygote.

4 The production of several hormones by the female co-ordinates the development of the egg and the wall of the uterus.

5 If fertilisation does not occur, the uterus sheds its lining in a process called menstruation. The entire cycle takes about 28 days.

6 If fertilisation does occur, the zygote travels down the Fallopian tube to the uterus.

7 Methods of contraception include the use of barriers, prevention of implantation, prevention of gamete production and regulation of times of sexual intercourse.

REPRODUCTION AND DEVELOPMENT

Questions for review

1 Of what advantage to a species is sexual reproduction?

2 List the parts that sperms pass through in the male reproductive system.

3 What is semen? What is its function?

4 How does the number of sperms produced by a man compare to the number of eggs produced by a woman? Explain the significance of this difference.

5 Describe the path of the egg in the female reproductive system.

6 Describe the effects of the sex hormones on the reproductive cycle in a woman.

7 Name and define four stages in the menstrual cycle.

Applying principles and concepts

1 Why must fertilisation occur in a Fallopian tube before the egg moves into the uterus?

2 Explain why a hormone imbalance that affects the pituitary gland might result in problems with ovarian function.

3 If an egg is fertilised, menstruation does not occur, and the lining of the uterus is prepared for implantation. How do hormones control this?

REPRODUCTION AND DEVELOPMENT

REPRODUCTION AND DEVELOPMENT

Growth and development

Learning Objectives

By the end of this chapter you should be able to:

- Understand the process of implantation
- Describe the early stages of development of the embryo
- Describe the development of the fetus
- Understand the functions of the placenta

- Describe the events leading to birth
- Describe the early stages of parental care
- Describe the changes taking place during puberty

Early development of the fertilised egg

Once fertilised, the egg begins the three to four day journey through the Fallopian tube to the uterus (Figure 2.2.1). Thirty-six hours after fertilisation, the single-celled zygote has become two cells. Two days later these cells have each divided twice more to give a microscopic ball of eight cells. In this condition, the egg completes its passage through the Fallopian tube and enters the uterus.

Four days after fertilisation, the egg is a cluster of 64 cells, which then begin to divide more rapidly. This stage corresponds to about day 20 of the menstrual cycle. This cluster of cells remains unattached for two days. The wall of the uterus at this stage is under the influence of the hormone, progesterone, and is becoming spongy and supplied with numerous small blood vessels. Twelve days after fertilisation the ball of cells has further increased in size and is now called an embryo. At this stage it becomes completely embedded in the wall of the uterus (endometrium). This is called **implantation**.

Growth of the fetus

As the embryo grows, it becomes surrounded by a thin, transparent membrane, the **amniotic sac**, containing fluid known as **amniotic fluid**. The embryo becomes suspended within the liquid by its **body stalk**, which attaches the embryo to the wall of the uterus.

By the sixth week of pregnancy, the **body stalk** has become longer and thicker and is called an **umbilical cord**. This cord contains the major veins and arteries that link the embryo to the mother.

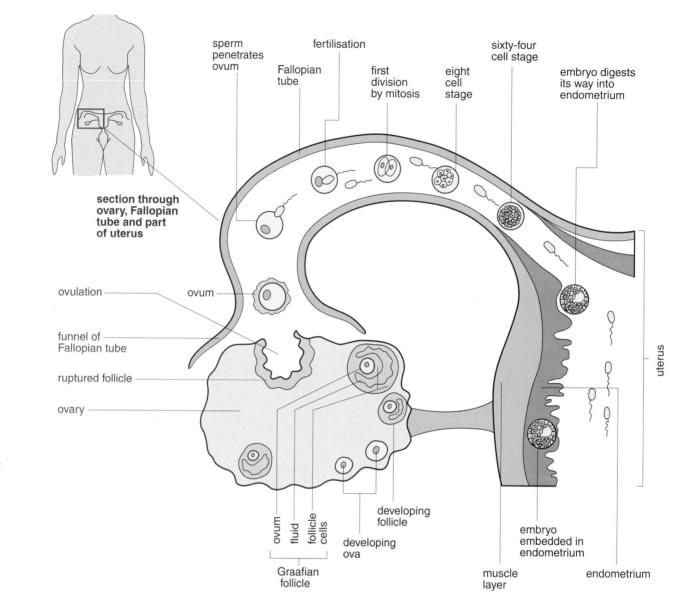

Figure 2.2.1 The fate of an egg

The **placenta** has also begun to develop at the end of the umbilical cord. This is a disc of tissue with finger-like projections which clings close to the wall of the uterus. It is through the placenta that exchange of food, oxygen and wastes takes place between the embryo's blood and the mother's blood (see Figure 2.2.2).

The embryo uses glucose, amino acids, minerals, vitamins and oxygen, brought to it by diffusion from its mother's blood. Carbon dioxide and urea are in turn diffused into the mother's blood and removed via her lungs or kidneys.

By eight weeks, the embryo is called a **fetus**. The first twelve weeks of fetal development are a particularly critical time. By twelve weeks after fertilisation all the organ systems of the fetus have formed. The heart pumps blood through the blood vessels, and the alimentary canal, excretory system and lungs are all present. The limbs, too, are formed, and although the mother cannot feel fetal movements at this stage, the fetus is already quite active.

The fetus, by this time, is recognisably human, and for the remainder of its time in the uterus,

fetus in the uterus

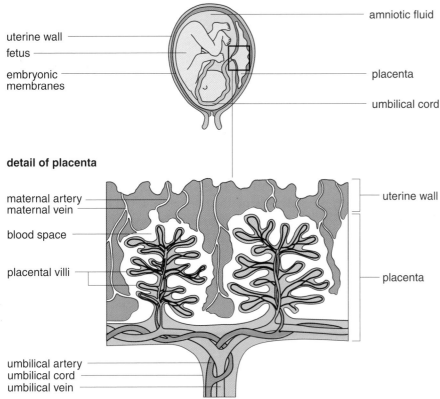

detail of placenta

Figure 2.2.2 The placenta

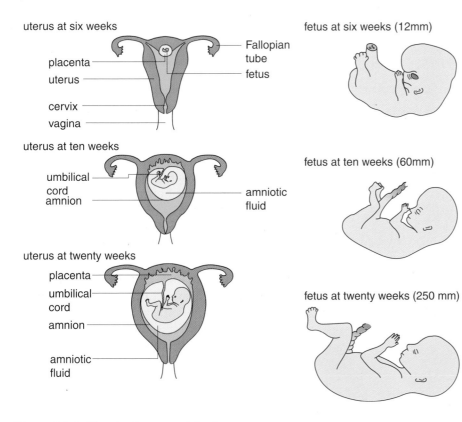

Figure 2.2.3 The development of the fetus

most of the changes consist of maturing and growth of all these systems and organs. Any damaging influence on the fetus can be particularly dangerous during the first twelve weeks. It may cause basic deformities which cannot be repaired in later development. An example of this was the tragic case of the use of the drug **thalidomide** by some pregnant women in the 1960s and 70s which led to damage occuring during limb formation.

At five months the fetus is 25 cm long, and the amniotic sac now contains a considerable volume of fluid. This prevents the fetus from damage resulting from physical pressure. It also allows the fetus to move about and helps keep a constant temperature (Figure 2.2.3).

By the end of a nine-month period, the fetus has usually turned head down (Figure 2.2.4).

It is now ready to be introduced to the outside world (Figure 2.2.5). In order to push the new baby out through the cervix of the uterus, the muscular layers of the uterus contract rhythmically in what is known as **labour**. The **breaking of the waters** is a general term used to describe the release of the amniotic fluid which surrounded the baby during its development.

This is an indication that labour is starting. **Contractions** are weak in the beginning but become stronger as birth proceeds. Eventually, the contractions force the fetus through the vagina or what is often referred to as the **birth canal**.

Once the baby has been born, the mother has to expel the placenta, or **afterbirth**. The umbilical cord runs through the vagina and is attached to the placenta, close to the wall of the uterus. Two firm ligatures are tied around the cord, a few centimetres away from the baby, and then a pair of sterile scissors is used to cut between the ligatures. If this is not done, there would be loss of blood when the cord is cut and the baby could bleed to death. Mild contractions of the wall of the uterus force the placenta out from the uterus and through the vagina.

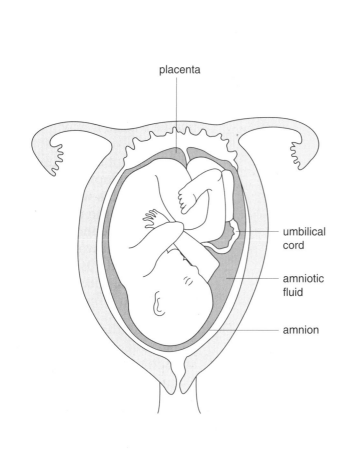

placenta

umbilical cord

amniotic fluid

amnion

Figure 2.2.4 The position of the fetus just before birth

Early parental care

All mammals are dependent on milk produced by the mother as soon as they are born. In humans, the milk produced by glands in the **breasts (mammary glands)** is different from cow's milk in that it contains less protein but more milk sugar (lactose), vitamins and mineral salts.

Breasts begin to develop at puberty but often remain relatively small and do not secrete milk. It is not until pregnancy that the breasts develop so that they are ready to produce milk when birth occurs. Early in pregnancy, there is a considerable growth of ducts and glandular tissue in the breasts, resulting in an increase in size. After birth, a hormone called **prolactin** is secreted from the pituitary and stimulates the production of milk by the glands. However, this hormone does not cause the milk to be released into the ducts, and so it is unavailable to the baby at this stage.

Another hormone, called **oxytocin**, causes muscle bands around the glands to contract, forcing milk into the ducts and towards the nipple. The release of this hormone in the mother is stimulated by the baby. As the baby sucks the nipple, nerve impulses are sent to the brain, which result in oxytocin being released from the pituitary. This hormone circulates in the bloodstream and soon reaches the breast, where it causes the muscle around the glands to contract, squeezing milk out of the milk glands and into the ducts of the breast. The milk in the ducts is then available when the baby sucks.

In some cases the baby may cause oxytocin to be released from the pituitary merely by its crying. The noise of the cries stimulates the mother's brain, again causing oxytocin to be released and milk to be forced out of the glands and through the nipple.

1 cervix starts to flatten

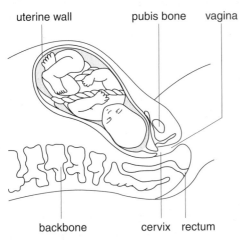

uterine wall   pubis bone   vagina

backbone   cervix   rectum

2 cervix flattens completely

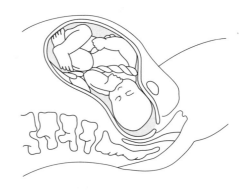

3 cervix partially opens

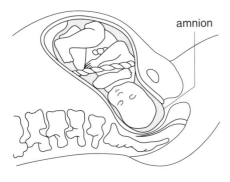

amnion

4 cervix fully opens and amnion breaks

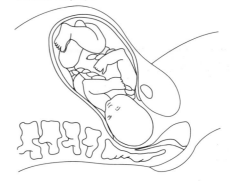

5 head rotates

6 head is born, shoulders and rest of body follow

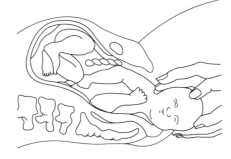

7 delivery of placenta

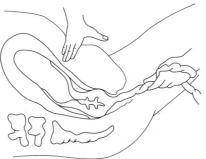

Figure 2.2.5 Birth

Puberty and hormones

The production of oestrogen and testosterone increases dramatically at the stage of growth called **puberty**. In girls this may occur at any time between 11 and 16 years, while in boys it usually occurs slightly later, between 13 and 17 years. The rise in hormone levels at this time of life is responsible for causing the development of the secondary sex characteristics. Important physiological changes also take place in both sexes at this time.

The changes are summarised in Table 1.

Table 1 Changes occuring to the body at puberty

Male	Female
Hair grows on face, chest, in pubic regions and armpits.	Hair grows in pubic regions and armpits.
Larynx enlarges and voice deepens.	Pelvis becomes broader, fat deposited on hips and thighs.
Body becomes more muscular.	Breasts develop.
The penis, scrotum, and prostate become larger.	Fallopian tubes, uterus and vagina enlarge, together with the linings of the uterus and vagina.
Sperm formation begins.	Ovulation and menstruation begin.
Feelings and sexual drives associated with adulthood begin to develop.	Feelings and sexual drives associated with adulthood begin to develop.

Questions for review

1 Describe the changes in the formation of the embryo from the zygote.

2 Describe the functions of the placenta.

3 Give an account of the ways in which hormones are related to milk production.

4 Describe the effects of hormones on growth during puberty.

Applying principles and concepts

1 Explain why it is important for both mother and child to have good pre-natal (before birth) care.

2 Compare the advantages and disadvantages of breast-feeding with feeding babies cow's milk.

Summary

1 If fertilisation occurs, the zygote travels down the Fallopian tube to the uterus.

2 The zygote becomes a many-celled sphere which becomes attached to the wall of the uterus.

3 The embryo becomes surrounded by an amniotic sac containing amniotic fluid.

4 The fetus is totally dependent on the mother for oxygen and nourishment. These materials diffuse through the thin membranes of the placenta but the blood of the fetus and mother do not mix.

5 The uterus contracts rhythmically in order to expel the fetus.

6 The mother's milk contains more lactose, vitamins and mineral salts than cow's milk.

7 The hormone, prolactin, causes milk production.

8 The hormone, oxytocin, causes muscles around the milk glands to force milk out of the nipple.

9 Puberty is due to a marked increase in testosterone and oestrogen.

10 At puberty secondary sexual characteristics become apparent.

REPRODUCTION AND DEVELOPMENT

REPRODUCTION AND DEVELOPMENT

Chromosomes and cell division

Learning Objectives

By the end of this chapter you should be able to:

- Understand the relationship between genes and chromosomes
- Describe the major stages in cell divison and understand their significance

- Understand the role of DNA and RNA in protein synthesis
- Describe how Mendel's theories apply to the behaviour of chromosomes and genes
- Explain sex determination

Sites of genetic material

In any study of biology we recognise variation and can explain it. However, we must consider its basic cause in more detail.

It would be ideal if we could look down a microscope and see the cause of variation revealed. Alas, this is not possible. Variations are handed down from one generation to the next through genetic material. Those cells which form a link between generations, the **gametes**, must contain genetic material and are therefore the best starting point for our study.

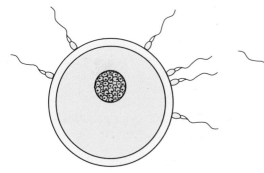

the sperms bump into the jelly coat around the egg

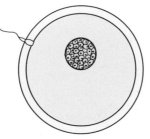

one of them penetrates the jelly

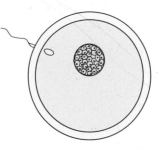

its head passes into the egg and the nuclei combine

Figure 2.3.1 Fertilisation

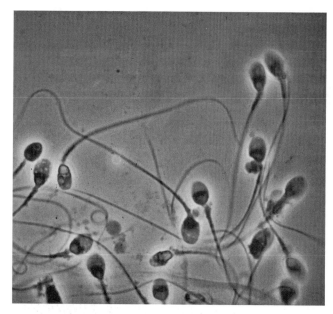

Sperms × 1344

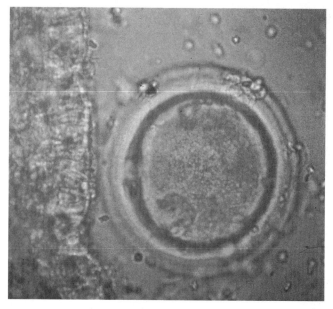

An egg in the Fallopian tube

The photograph (above left) shows a male gamete (**sperm**). In common with other cells, it has a nucleus and a membrane. Unlike the female gamete (**egg**), which you can see in the photograph (above right), it has very little cytoplasm. Male gametes contain all of the genetic material in the nucleus, which is the so-called 'head' of the sperm. Sperms are **motile**

(they can move). When fertilisation takes place the nucleus enters the female gamete and fuses with the female nucleus (Figure 2.3.1).

The nucleus holds all the secrets of genetics. It contains all of the chemicals which are used to transfer genetic information from one individual to its offspring.

Cell division

The cell division that you will see in your investigation is called mitosis. Apart from the special case of gamete formation, most cell divisions include mitotic division of chromosomes.

The photograph (right) shows a lengthwise section of a root tip; different cells at different stages in mitosis can clearly be seen. The stages in mitosis are shown diagrammatically in Figure 2.3.2.

The nucleus divides in four main stages. Prior to mitosis, each chromosome makes a copy of itself. At the start of mitosis, the original and the copy separate and migrate to opposite poles along a **spindle**. Two **daughter nuclei** are thus formed. Division of the cytoplasm and the formation of a new cell wall then occurs.

Mitosis enables each new cell (**the daughter cell**) to have an identical set of chromosomes to

Mitosis in plant cells. You can see several stages going on at once

A closer look at the nucleus

The best material to use is the growing region of a root tip of garlic. This plant is available throughout the year and is easy to grow. Simply place the garlic bulb on the top of a test tube containing water. In a few days roots will appear.

Procedure

You will be given root tips which have been stained with Feulgen's stain - this reacts with chemicals in the nucleus to produce a pink colour.

1 Cut 2 mm off the end of the root tip using a sharp scalpel.

2 Place the tip on a microscope slide in a drop of 45% acetic acid. Use two mounted needles to pull the root tip apart. Try to break down the tissue into very small parts.

3 Place a cover slip on the material. There should be just enough acetic acid to fill out the cover slip.

4 With the slide resting on some filter paper, put a few layers of filter paper on top of it. Press your thumb straight down on the region of the cover slip. Avoid any sideways pressure. This should flatten the cells and separate the genetic material (chromosomes).

5 After about five seconds, peel off the blotting paper. The acetic acid should fill the space under the cover slip.

6 Examine the slide under the microscope. Use the high power objective after focusing under low power. Draw the cells.

You should be able to make out the chromosomes. Chromosomes contain genes – the carriers of genetic information – which can be seen in the photograph (below). Try to find the earliest stage at which the chromosomes are visible as double structures.

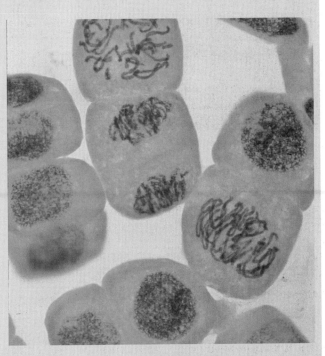

Photomicrograph of a plant cells showing chromosomes

Look at several cells and compare them to ·Figure 2.3.2. As a result of your observations you should be aware of the following:

a The nucleus is the most likely site of the genetic material.

b The nucleus can contain chromosomes.

c Chromosomes are only visible in cells which are actively dividing.

the original cell. It is the means by which all the cells of an organism are derived from a zygote. It also allows organisms to reproduce asexually. Therefore, all asexually-produced offspring from one organism have the same genetic information as that of their parent.

Sexually-produced offspring arise from the fusion of male and female cells. The production of these cells takes place by a special form of cell division called **meiosis**.

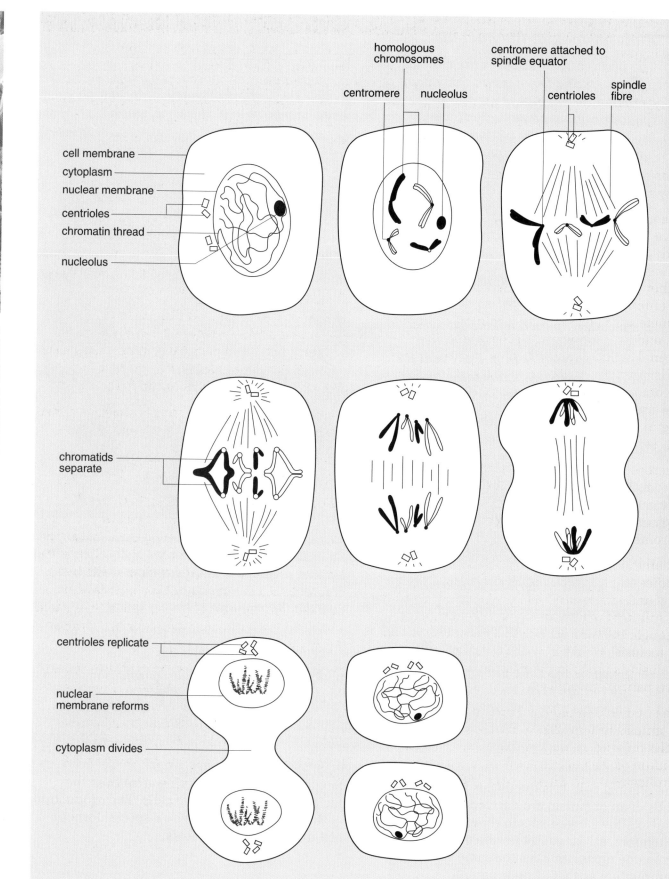

Figure 2.3.2 Mitosis

Sex cells and fertilisation

The largest known cells are eggs. They contain little but cytoplasm; their bulk is mostly water and stored food. The important part of the egg is the nucleus. Sperm cells are hundreds or even thousands of times smaller than eggs. They are little more than a nucleus attached to a vigorous 'tail'. When a sperm finds its goal, its nucleus joins the egg nucleus to form the nucleus of a new and separate cell.

The nuclei of both kinds of sex cells generally have fewer chromosomes than other cells in the body of the same organism. The reason for this becomes clear when we see what would happen if they did not. Cells in the human body have 46 chromosomes. If each matured sex cell, or gamete, also had 46, then a baby would have 92, and its children would have 184. Yet all normal cells in human bodies have 46. Human cells normally contain 23 pairs of chromosomes. One set comes from the mother's egg and the other set comes from the father's sperm. Mitosis ensures that each new cell gets a full set of chromosome pairs.

Microscopic studies of developing human sperms and eggs show that they have only 23 chromosomes, one from each pair. How does an organism produce cells with half the number of chromosome pairs?

Of all the countless millions of cells in our bodies – and those of other organisms – only egg- and sperm-producing cells divide in a way to split up the chromosome pairs. Logically enough, the process is called **reduction-division** or **meiosis**. To see how it works we can follow meiosis as it occurs in an animal that has only two pairs of chromosomes.

The first step in meiosis is similar in some ways to mitosis (Figure 2.3.3). Each chromosome pairs off with its opposite number across the middle of the nucleus, and each chromosome duplicates itself. There is a double chromosome for each original, the halves of which are **chromatids**. Now, just as in mitosis, the nuclear membrane disappears, but unlike mitosis each double member of a pair goes to the new cells. At this point we have two cells, each containing two double chromosomes – one from each original pair.

A brief resting period follows, then begins a new wave of nuclear events, during which double chromosomes break apart and each chromatid becomes a separate **chromosome**. Then the cells divide again. There are now four cells, each with two chromosomes, one of each pair of chromosomes in the original.

The process of meiosis is illustrated diagrammatically in Figure 2.3.4. Both sperm and egg cells undergo meiosis. When the sperm unites with the egg, each provides half the chromosomes for the new individual.

A cell with the full set of chromosomes is called **diploid**.

A cell with half the normal number of chromosomes is called **haploid**.

Since it is highly unlikely that there ever existed a cell with this identical set of chromosomes, each fertilised egg cell is unique. The advantage of sexual reproduction lies in the fertilised egg which makes each new life just a little different from either of its parents. This variation among offspring may produce one that may be able to adapt to changing conditions in the environment.

Once the egg has been fertilised, all further cell divisions produce cells with the full number of chromosomes. Eventually the new organism will reach maturity and its time will come to reproduce. Its reproductive organs will then produce sperms or eggs and the cycle of life will have come full circle.

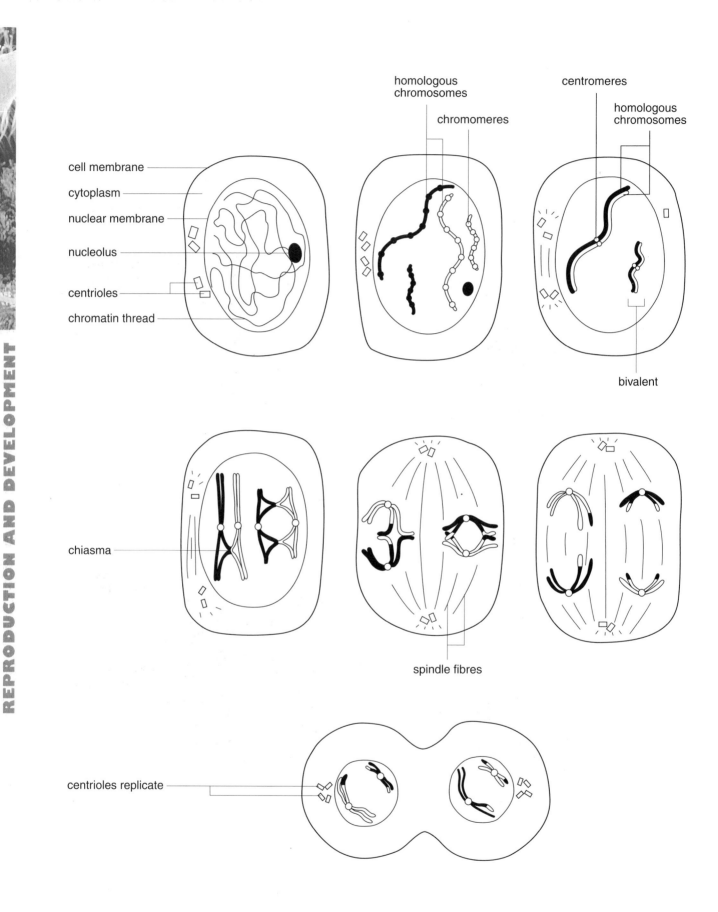

Figure 2.3.3 First division of meiosis

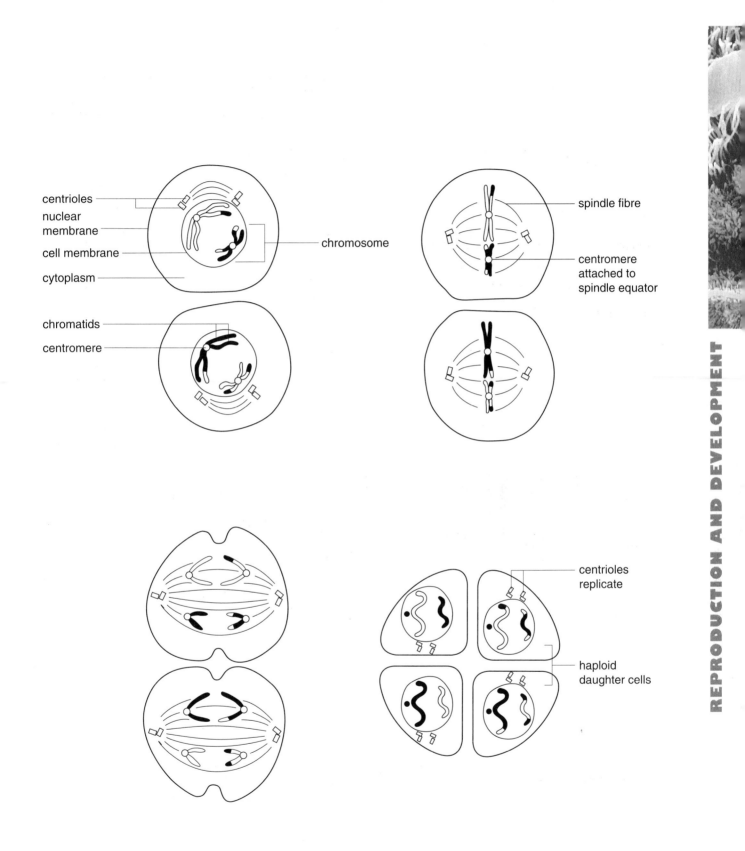

centrioles

nuclear membrane

cell membrane

cytoplasm

chromosome

chromatids

centromere

spindle fibre

centromere attached to spindle equator

centrioles replicate

haploid daughter cells

Figure 2.3.4 Second division of meiosis

REPRODUCTION AND DEVELOPMENT

Chromosomes, genes and DNA

An organism's characteristics are passed down from generation to generation by genes. But what are genes, where can they be found and what do they actually do?

Life is a series of chemical reactions. It is not surprising, therefore, that quite early in the history of genetics, scientists knew that genes must be chemicals. Proteins are the most essential chemicals found in living cells. The early geneticists guessed that genes probably existed as protein contained in the chromosomes in the nuclei of cells.

By 1950, it became clear that it was not the protein in chromosomes that passed on the code of life from generation to generation. It was another component of cells: **nucleic acids**. These are some of the largest and by far the most fascinating of all life's molecules. Two forms are known:

1 **deoxyribonucleic acid (DNA)**, which is found in all chromosomes

2 **ribonucleic acid (RNA)**, which is found in the cytoplasm and nuclei of cells.

It was DNA that carried the genetic code. The structure of the DNA molecule was discovered in 1953 by the American, James Watson, and the English scientist, Francis Crick, working at the Cavendish Laboratory in Cambridge.

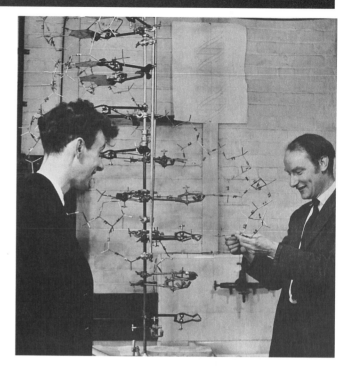

James Watson (born 1928) and Francis Crick (born 1916) with their first model of DNA, made in 1953

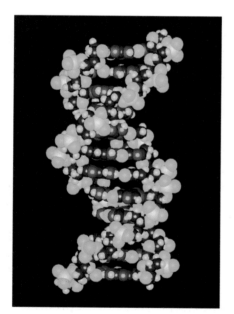

A computer model of DNA

Nucleic acids, like proteins, are made of many units strung together. DNA has a ladder-like structure of two long **sugar-phosphate chains** joined together by connecting **bases** (the 'rungs'). The ladder is twisted to form a three-dimensional **double helix** as shown in the photograph (left). Figure 2.3.5 shows a model of a strand of DNA.

Watson and Crick discoved that the rungs of the ladder are made of four different types of bases: **guanine**, **cytosine**, **thymine**, and **adenine**. The bases fit together as shown in Figure 2.3.5. Guanine only pairs with cytosine, and thymine only pairs with adenine. The differences between one DNA molecule and another – or one gene and another – depends on the pattern of these base pairs. This is how genes produce certain effects in organisms. For example, the order of base pairs which produces blue eyes is different from the order of base pairs which produces brown eyes.

There is an almost unlimited number of possible arrangements of base pairs, if you consider that a strand of DNA can be over

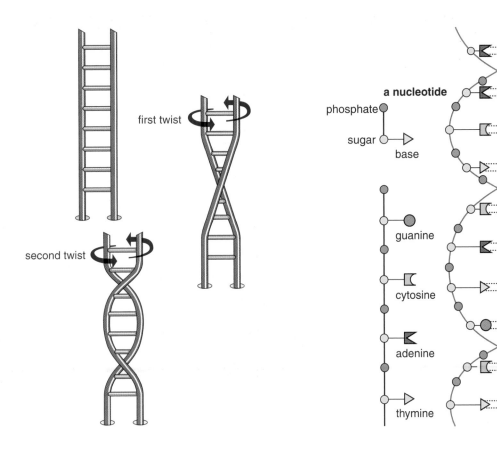

Figure 2.3.5 The structure of DNA

10 000 base units long. Therefore, there is an almost unlimited number of possible genes in plants and animals. The DNA contains the genetic code which tells an organism how to develop.

How do genes work?

How do genes control the development of an organism? How can such microscopic particles have such a staggering effect on life?

Genes consist of DNA. DNA expresses a code that determines which chemical reactions take place in a cell and at what speed. It does this by determining which proteins are made (**synthesised**) in the cell. The growth and development of a cell is determined by the type and speed of the chemical reactions taking place within it. So, by controlling protein synthesis, DNA controls the life of the cell, and hence the development of the organism.

How does DNA control protein synthesis?

Proteins are made of building blocks called **amino acids** (see page 17). The amino acids are linked together in chains. The different ways in which different amino acids are linked together determines the type of protein synthesised (Figure 2.3.6). DNA is able to regulate how the amino acids are arranged. The types and arrangement of the bases in the DNA molecule act as a code that determines which amino acids are linked together. So, by determining the form and arrangement of these basic building blocks, DNA controls protein synthesis.

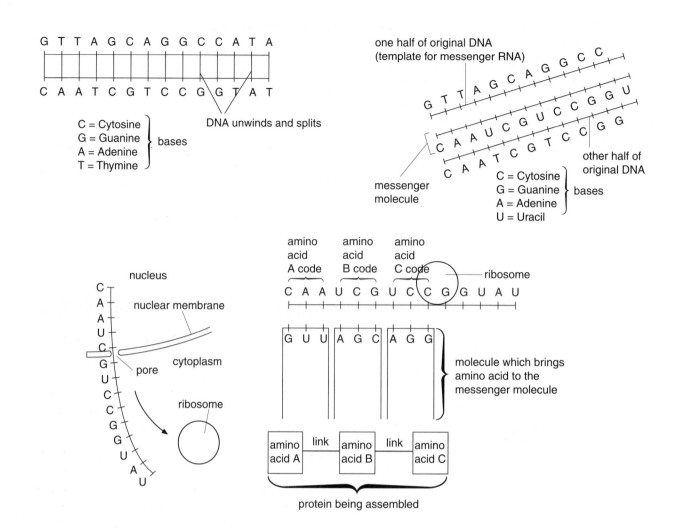

Figure 2.3.6 Protein synthesis

The following is a very simplified account of how DNA is used for making proteins:

1   The long molecule of DNA (remember it is like a twisted ladder) unwinds and splits along its length between the bases.

2   One half of the molecule now acts as a pattern for the formation of a **messenger** molecule. This consists of **RNA** and is made by new bases found as a mixture in the nucleus. These strings of RNA line up opposite their partners on the original half of the DNA. In this way, they form a **single strand**. The result is that the code originally present in the DNA is now also on the messenger molecule.

3   The messenger molecule (**transfer RNA**) then passes through the nuclear membrane to structures in the cell called **ribosomes**.

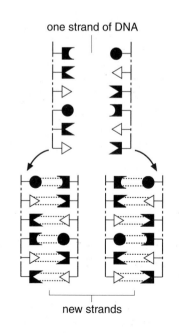

Figure 2.3.7 DNA replication

4 The code on the messenger molecule then determines which amino acids from the cytoplasm of the cell become linked together and therefore the type of protein synthesised.

The discovery that DNA carries the code of life was an enormous step forward in molecular biology. The next step was the discovery of how the molecule replicates itself.

DNA replication

Remember that before a cell divides, the chromosomes duplicate themselves. As we have just learned, the chromosome actually consists of a long chain of DNA – it is this that duplicates itself.

The two long strands of DNA separate and then the free bases present in the cell nucleus align themselves with the separated strands (guanine with cytosine, thymine with adenine). This results in the creation of two new double helices (Figure 2.3.7).

Sex chromosomes

TH Morgan, who discovered sex chromosomes

The discovery of special chromosomes which determined the sex of an organism took place in 1910. An American geneticist, T H Morgan, was studying the genetics of the fruit fly, *Drosophila melanogaster*.

You may have seen this small fly hovering around over-ripe fruit (especially bananas), sometimes in vast numbers. Just like Gregor

Mendel's significant discoveries were made using very common, but perhaps unlikely, organisms (see page 146), so Morgan's discoveries were equally important. Who would have thought that the humble fruit fly would have played such a major role in science?

Among thousands of red-eyed flies raised under laboratory conditions, Morgan found one mutant fly with white eyes. This fly was a male and was mated with a normal red-eyed female. The F1 generation consisted entirely of red-eyed flies. By following Mendel's logic, this meant that the gene for white eyes was recessive to the gene for red eyes. Next, members of the F1 were mated to produce the

Male and female fruit flies, *Drosophila melanogaster*

F2 generation. The results agreed with Mendel's observations on peas; i.e. a 3:1 ratio of red-eyed flies to white-eyed flies. However, Morgan noticed that all of the white-eyed flies were males! The gene for eye colour was in some way linked to the sex of the animal. White eye colour is controlled by a **sex-linked gene**.

A clue to the solution of this puzzle was seen in the chromosomes. It was already known that there is a difference between the chromosomes of male and female *Drosophila*. Of the four pairs of chromosomes in each cell, three pairs are identical in males and females. The other pair is different. The straight, rod-shaped chromosomes of this pair are called X chromosomes. The hook-shaped member of this pair (found in the male only) is called the Y chromosome (Figure 2.3.8).

When combined with a knowledge of meiosis, these observations led to an important conclusion. Males produce two different kinds of sperms so far as these chromosomes are concerned. Half the sperms would carry an X chromosome and the other half would carry the Y chromosome. Females, on the other hand, would produce only one kind of egg – with one X chromosome. Both eggs and sperms, of course, carry one chromosome of each of the other pairs.

Because of their connection with the sex of the flies, the X and Y chromosomes are called **sex chromosomes**. All other chromosomes are called **autosomes**. The discovery of two kinds of sex chromosomes suggested that perhaps they would provide an explanation for the determination of sex. Flies with two X chromosomes are always females. Those with

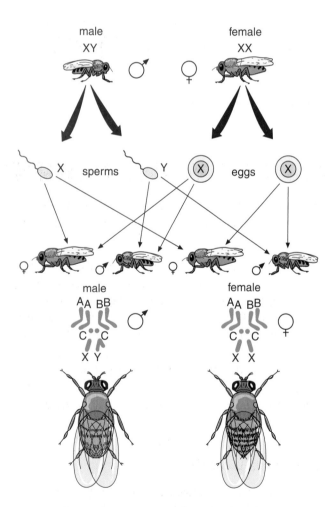

Figure 2.3.8 X and Y chromosomes in fruit flies

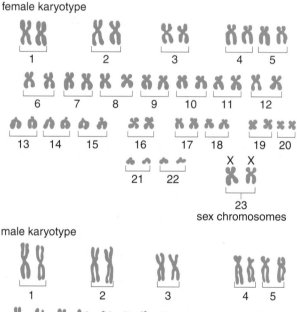

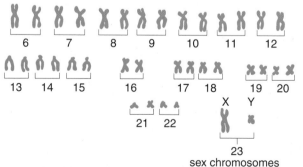

Figure 2.3.9 The human karyotype

an X and a Y are always males. The sex of a fly then depends if the egg (with an X chromosome) is fertilised by a sperm with an X chromosome or one with a Y chromosome. So it is the male fly's sex cell that determines the sex of the offspring.

Since the time of Morgan, experimental and research techniques have been improved. It is now possible to show that a similar difference in one pair of chromosomes is common among all animals. (Most plants, however, do not have separate sexes. But those with separate male and female individuals have sex chromosomes.) The human pattern of sex chromosomes is similar to that of the fruit fly.

Humans have 23 pairs of chromosomes (Figure 2.3.9). Twenty-two pairs of these are autosomes (non-sex chromosomes) and one pair consists of the sex chromosomes. In men, the Y chromosome is very small, compared to the X chromosome. We can explain sex determination in humans the same way as we do in fruit flies.

(Note: not all animals have sex-determining sperms. Birds and butterflies have males with XX chromosomes and females with XY chromosomes.)

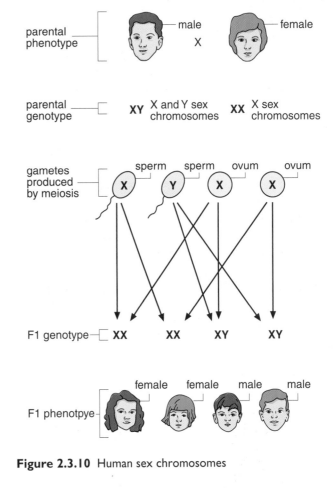

Figure 2.3.10 Human sex chromosomes

What is inheritance?

We have seen that the bridge that connects one generation to the next is microscopic. It consists of an egg cell and a sperm cell. Within these tiny bits of living matter are the plans for the next generation. The sperm fertilises the egg and then one of the most remarkable series of changes known to science begins. The instructions within the human cell control the development of the embryo into a human rather than into an elephant or a mouse. This is the case for every other organism that reproduces sexually.

The instructions are in the form of **chemical messages**. The way in which these messages are passed from one generation to the next is **inheritance**. The study of inheritance has become a branch of science called **genetics**. Today most people have heard of the term 'genetics'. Genetic engineering, genetic counselling, and genetically inherited diseases are terms which are used in newspapers, radio and TV.

We are probably much more interested in the medical aspects of genetics than anything else. This is not surprising because we hear of so many disorders that can be passed from parents to their children through genetic inheritance. If you are a farmer or a gardener you would certainly be interested in knowing about the methods of passing useful characters from one generation to the next in selective breeding.

Selective breeding

For thousands of years the selected pedigree animal has been prized. Such animals have had a recorded ancestry for many generations.

A pedigree dairy herd

Grain yield can be increased by selective breeding

The dairyman wants selected pedigree cattle from which to breed his milking cows. The hunter wants a pedigree dog. Similarly, pure lines of cultivated plants are needed by the farmer.

Characters such as grain yield, fruit yield and disease resistance are desirable qualities that can be maintained through selective breeding. Although humans have learned a great deal about selective breeding, we do not pretend to know all the solutions to our problems. Some attempts have had little or no success. We know that selective breeding sometimes gives the results we want, but sometimes it does not.

We have known that inheritance was involved in selective breeding for many years. How it operates remained a mystery until the nineteenth century.

It might seem strange to you now, but the whole science of genetics really began with a monk experimenting with pea seeds in 1865. Using this unlikely material and a great deal of patience, this Augustinian monk, Gregor Mendel, set out the first principles of inheritance. For his unique contribution to science, he is often called the 'Father of Genetics'.

Gregor Mendel (1822–1884)

The work of Mendel

Although Mendel probably performed very few experiments that had not been done before, he succeeded where others had failed. Possibly his success was due to the unusual combination of skills he brought to the task. He was trained in mathematics as well as in biology. With this background, he planned experiments that at the time were novel in three respects:

1 Instead of studying a small number of offspring from one mating, Mendel used many identical matings. As a result he had enormous numbers of offspring to study.

2 As a result of having large numbers to study, he was able to apply statistical methods to analyse the results.

3 He limited each cross to a single difference, a single pair of contrasted characters at a time.

People who had carried out genetic crosses with plants and animals before Mendel did not concentrate on one character at a time. **Pure-breeding** stock was not always used (pure-breeding individuals receive similar genes from both parents). Characters were often masked by each other.

Mendel selected garden peas for his experiments because he knew they possess many varieties. The plants are easy to cultivate and cross, and the generation time is reasonably short. Finally, and of great importance, the plants are self-pollinating. The significance of this requires an explanation.

Let us consider the structure of the pea flower (Figure 2.3.11) and see how fertilisation usually takes place. Pollen from the anthers falls onto the stigma of the same flower. This happens before the flower opens fully. Pollen tubes develop from the pollen grains. These carry the male gamete to the female gamete in the ovule.

If you want to pollinate one pea flower with pollen from another plant, you must remove the

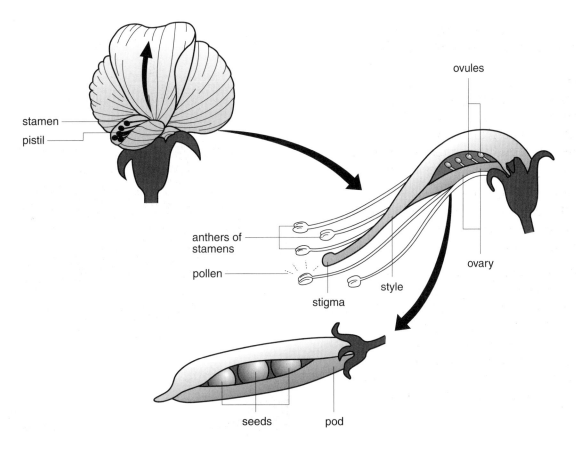

Figure 2.3.11 The pea flower

anthers from the flower before its own pollen is mature. Later, when the stigma is ready to receive pollen, you can dust it with pollen taken from some other flower of your choice. In this way the parents of the next generation can be controlled.

Mendel made sure that the plants he started with were all pure-bred for each character he was studying. He did this by letting the plants fertilise themselves for a number of generations. The offspring of each generation were studied to make sure they were all like one another and like the parent plant. Then Mendel made hundreds of crosses by dusting the pollen of one kind of plant on the stigmas of plants of another kind. For example, he pollinated plants from a type whose seeds were always round, with pollen from a type whose seeds were always wrinkled.

In every case he found that all the offspring resembled one of the parents and showed no sign of the character of the other parent. Thus all the crosses between plants with round seeds and plants with wrinkled seeds produced offspring whose seeds were always round. One character seemed to 'dominate' the other. Mendel therefore called this character the dominant character.

Table I Dominance in seven pairs of characters in garden peas

Character	Dominance
seed shape	round seed dominant to wrinkled seed
seed colour	yellow seed dominant to green seed
seed coat colour	coloured seed coat dominant to white seed coat
pod shape	inflated pod dominant to wrinkled pod
pod colour	green pod dominant to yellow pod
flower position	axial flowers dominant to terminal flowers
stem length	long stem dominant to short stem

Mendel's results – a summary

Mendel used seven pairs of contrasted characters (see Table 1).

(Note: Axial flowers occur in the angles between stems and branches.)

When a cross is made between two plants, pure-bred for contrasted characters, (e.g. round seed

Table 2 Mendel's results with two generations of garden peas

Characters selected	FI	Selfed	F2	Ratio
round × wrinkled seeds	all round	round × round	5474 round 1858 wrinkled	2.96:1
yellow × green seeds	all yellow	yellow x yellow	6022 yellow 2001 green	3.01:1
coloured × white seed coat	all coloured	coloured × coloured	705 coloured 224 white	3.15:1
inflated × wrinkled pods	all inflated	inflated × inflated	882 inflated 299 wrinkled	2.95:1
green × yellow pods	all green	green × green	428 green 152 yellow	2.82:1
axial x terminal flowers	all axial	axial × axial	651 axial 207 terminal	3.14:1
long × short stems	all long	long × long	787 long 277 short	2.84:1

versus wrinkled seed), the parents are the **P1 generation**. The offspring are the **first filial generation F1**. These symbols will help us follow Mendel's steps in his thinking.

He let the F1 plants pollinate themselves. This produced the **F2 generation**. The dominant character appeared in 75% of the F2 generation, while in 25% the other character reappeared.

Lise Merehall

Since the other character had receded into the background for a generation, he called it **recessive**. He also noted no in-between forms. The new seeds were either round or wrinkled, or tall or short, etc., just like the Pl.

Mendel's actual results are shown in Table 2. The 3:1 ratio is obvious in all cases.

Inheritance of a single character (monohybrid inheritance)

Mendel made his greatest contribution to genetics by explaining his observations. He began by using symbols to represent the characters he was dealing with. His mathematical training encouraged him to use symbols rather than written descriptions.

He assumed that the character for tall plants was caused by a dominant factor. He used a capital T ('big T') to symbolise this element. The character for short plants, the only alternative to tall plants, was caused by a recessive factor, t ('little T'). Basically we use the same idea today.

In 1910, many years after Mendel's work, the genetic factor was given the name **gene**. Next, Mendel assumed that every plant had a pair of genes for each character. These pairs of genes are called **allelomorphic pairs** or **alleles**. An allele is one of two alternative forms of a gene which may occupy the same part of a chromosome. For example, if T represents tall and t represents short, TT or Tt or tt can be the allelomorphic pair. T and t are at the same place on corresponding pairs of chromosomes.

Mendel was convinced of the existence of alleles because some parent plants with the dominant gene produced some offspring with the recessive gene. Therefore, every F1 plant must have each type of gene. The Fl plant could be represented by Tt. Mendel also assigned a pair of genes per character even to plants which are true-breeding. A plant from parents that bred true for tall plants was therefore TT. Similarly, a plant from parents that bred true for short plants was tt. These paired symbols representing the genes of an organism are its **genotype**. When an organism has identical

factors in its genotype (e.g. TT or tt), it is known as **homozygous**; if it has different factors (e.g. Tt), it is known as **heterozygous**.

Using this logic Mendel was then able to test his hypothesis about genes. If he knew the genotype of each parent, he could predict the kinds and proportions of gametes each parent could produce. From this, he could predict the kind and proportions of the offspring.

If every plant had a pair of genes for each character, was there any rule about how these genes were passed on to the next generation? Mendel thought about the short plants that appeared in the F2 generation. These could not carry the dominant gene T. They must have received the recessive t from the F1. Remember the F2 was produced by self-pollinating the F1.

The next question was: how frequently do gametes that carry t occur among all the gametes produced by the F1 Tt parents? Mendel reasoned back from the proportions of tt short plants in the F2. These amounted to one-quarter of the entire F2 generation. So the frequency of t in the eggs and male gametes produced by the F1 generation should be one half (since one half multiplied by one half = one quarter). This means that half the gametes of a Tt plant would carry T; the other half t.

Figure 2.3.12 shows Mendel's classic experiments with peas. If a pure-breeding tall variety was crossed with a pure-breeding short variety all the offspring were tall. When two of these plants were crossed, three tall offspring were produced for every short one. Mendel saw that these results could be explained if the characteristics were inherited as 'particles'.

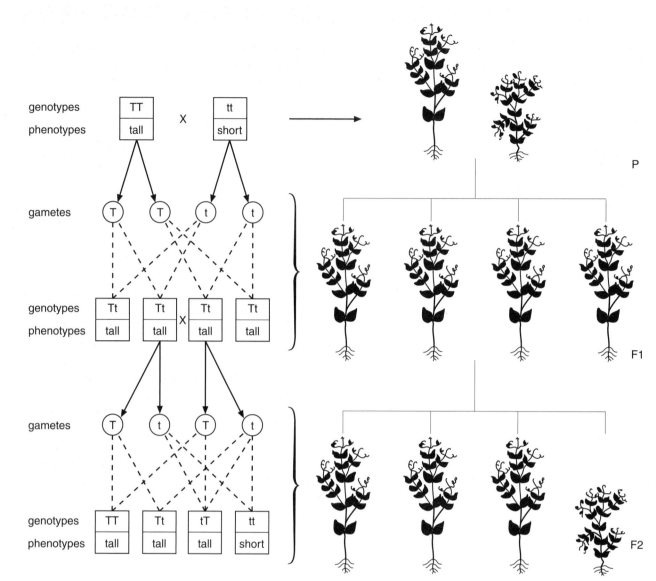

Figure 2.3.12 Mendel's results

Each plant had two of these particles, and one was dominant to the other. In this case the tall character (T) was dominant to short (t).

Mendel thus arrived at a general rule – **the law of segregation**:

> 'The two members of each pair of genes must separate when gametes are formed, and only one of each pair can go to a gamete'.

If a parent is TT its gametes will all inherit one or other of its T genes, but not both. If the parent is tt, its gametes will all inherit one of its t genes. If the parent is Tt, half the gametes will inherit its T gene, and the other will inherit its t gene.

An exception to Mendel's rule

Mendel dealt only with genes which showed either complete dominance or complete recessiveness. Blending of characters upset his 3:1 ratio found in F2. We know, however, that Mendel knew that some plants appear to show blending. On page 37, there is a description of the ABO blood grouping system. These blood groups are controlled by allelles but the alleles for group A and for group B have equal dominance (co-dominance) and both are dominant over the allele for group O. It is an example of multiple gene inheritance. The inheritance of a particular blood group is determined by a gene that controls the production of a specific antigen (see page 226):

- People with group A blood have the gene that produces the a antigen.
- People with group B blood have the gene that produces the b antigen.
- People with group AB blood have the genes that produce both antigens.
- People with group O blood have neither gene.

The following table illustrates inheritance of blood groups and shows the possible groups produced from parents of different genotypes. If we know whether the parents are homozygous or heterozygous, we can determine the possible blood groups of their children.

Consider a father with blood group O and a mother with blood group A. The possible number of blood groups of the children is then quite small:

Possible genotypes of the parents	Possible blood groups of the children
mother AA or AO father OO	A and O

Sometimes an understanding of blood group inheritance is used in Courts of Law. In a paternity suit, a woman may allege that a particular man is the father of her child. The woman's and man's blood can be tested and may help to withold or disprove the allegation. Using the last example, if the child has A or O type blood, the father cited in the case may indeed be the father. However, if the child's blood type was B or AB, this man could not possibly be the father.

Table 1 The inheritance of blood groups

Phenotypes (blood groups)	Group A	Group B	Group AB	Group O
possible genotypes	AA or AO	BB or BO	AB	OO
Example cross				
parental blood groups	possible genotypes			
mother (B) × father (A) genotypes	BB × AA | B | B A | AB | AB A | AB | AB	BO × AO | B | O A | AB | AO O | BO | OO	BO × AA | B | O A | AB | AO A | AB | AO	BB × AO | B | B A | AB | AB O | BO | BO
phenotypes	100% AB	25% AB 25% A 25% B 25% O	50% AB 50% A	50% AB 50% B

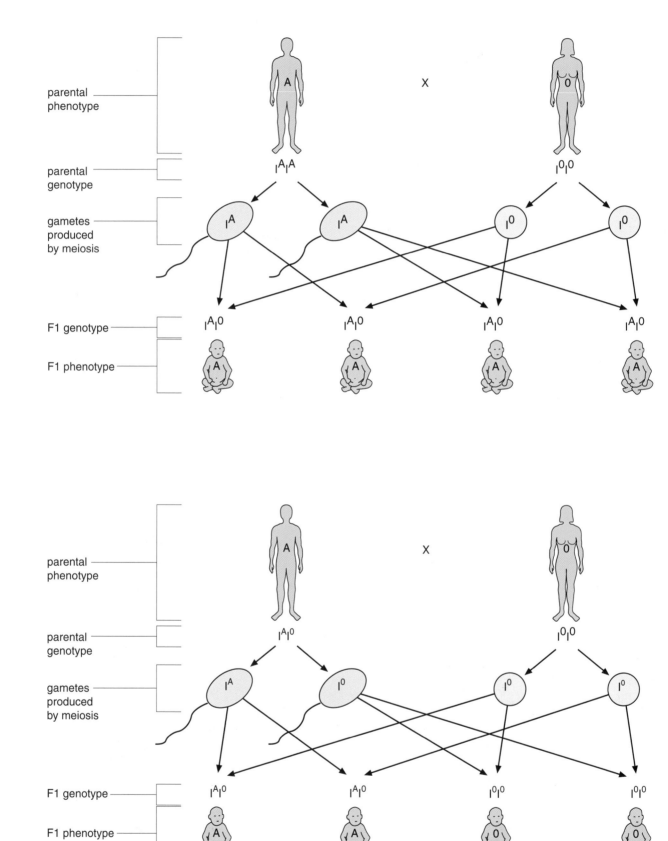

Figure 2.3.13 Blood group genetics

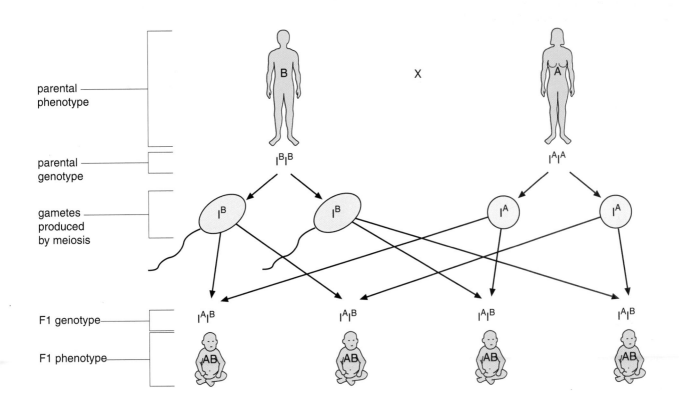

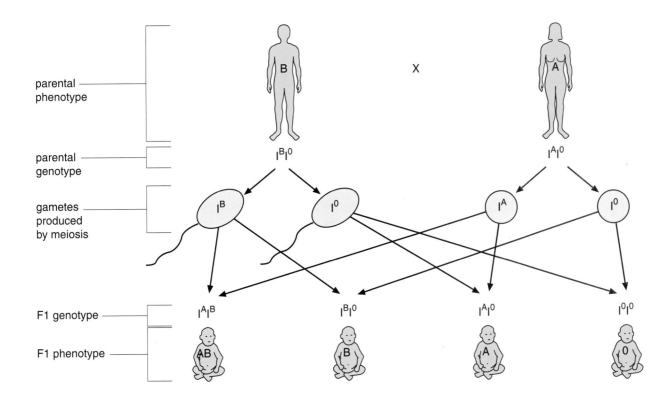

Figure 2.3.13 continued

Summary

1 Organisms grow by a type of cell division called mitosis.

2 Each daughter cell produced in mitosis has the diploid chromosome number and the chromosomes of the daughter cells are identical to those of the parent cell.

3 All body cells of members of the same species contain the same kind and number of chromosomes.

4 Sperms and eggs are produced by meiosis and are haploid.

5 When sperms and eggs join at fertilisation, the diploid number is restored.

6 The synthesis of proteins is one of the most important processes in living cells.

7 DNA controls protein synthesis through a code of bases. By transcription, this code is passed on to messenger RNA.

8 Messenger RNA carries the code into the cytoplasm where the messenger RNA acts as a template for the building of proteins.

9 Units of another type of RNA (transfer RNA) bring amino acids to the template where a ribosome reads the code and a protein is assembled from the amino acids.

10 More than a century has passed since Gregor Mendel's experiments with garden peas led to the formulation of the basic laws of genetics.

11 Modern genetics is answering many questions about human heredity.

12 Most characteristics that we inherit are helpful, but some are harmful.

Questions for review

1 What does 'genetics' mean?

2 What were the advantages to Mendel of using peas for his experiments?

3 State Mendel's principle of dominance.

4 Distinguish between 'genotype' and 'phenotype'.

5 What is meant by co-dominance?

Applying principles and concepts

1 Gregor Mendel formulated his first law of genetics which states 'Of a pair of contrasted characters, only one can be represented in a gamete by its germinal unit'.
  a) Give the modern name for 'germinal unit'.
  b) State where these germinal units are found in the gametes.
  c) A red-haired woman marries a brown-haired man, and all the children are brown-haired. Explain this genetically.

2

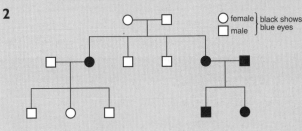

Figure 2.3.14

Figure 2.3.14 is part of a family tree showing the distribution of brown eyes and blue eyes.
  a) Which number represents a man who is homozygous for blue eyes?
  b) Which number represents a woman who must be heterozygous?
  c) Which of the eye colours is controlled by a dominant gene?

  d) Which part of the family tree enables you to answer part c)?

3 In humans, the gene for tongue rolling, R, is dominant to the gene for the inability to roll the tongue, r.
  a) Using these symbols, give the genotypes of the children that could be born from a marriage between a heterozygous father and a non-tongue-rolling mother.
  b) State whether the children would be tongue-rollers or not.

4 A boy was interested to see how eye colour was inherited in his family, so he wrote down the colour of the eyes of all his family. The information and his deductions are given in Figure 2.3.15. Unfortunately, he could not finish it.

Fill in the gaps after copying the table.
  a) Write the correct words for the numbers 1–17.
  b) From the results, state the dominant eye colour.
  c) Also from the results, state what recessive means.
  d) How many factors for eye colour do the gametes shown contain?

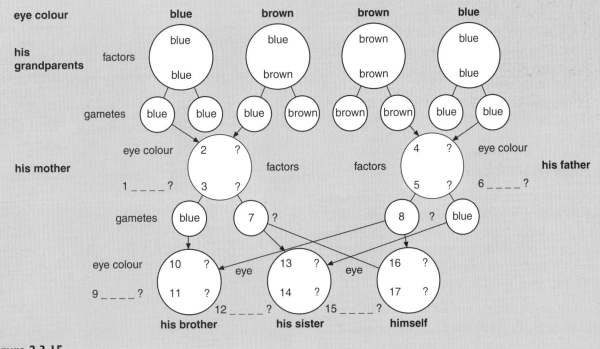

Figure 2.3.15

REPRODUCTION AND DEVELOPMENT

Variations – their causes and significance

Learning Objectives

By the end of this chapter you should be able to:

- Understand the difference between continuous and discontinuous variation
- Understand that variation leads to evolution through natural selection
- Understand the major causes of variation

- Appreciate the effects of radiation in terms of mutation
- Realise the effects of radiation on human health

Variation

One of the most fascinating things about living organisms is their enormous **variety**. They differ in appearance, behaviour, and where they live. A garden snail, a cockle, a lion, a horse and a zebra are all animals. They all differ from each other. However, the garden snail and the cockle can be grouped together and the lion, horse and zebra can be grouped together. But the zebra and a horse resemble each other more than either resembles a lion.

So living organisms can be divided into **groups**, the members of which have similar features. Inside each large group, smaller groups can be made. Let us consider ourselves and see how we are different from most other organisms but how we are also similar to some.

Right at the end of this branched, tree-like system of classification we find humans as a distinct species of animal. A species is defined as a group of individuals which can interbreed

The human race consists of many different ethnic groups

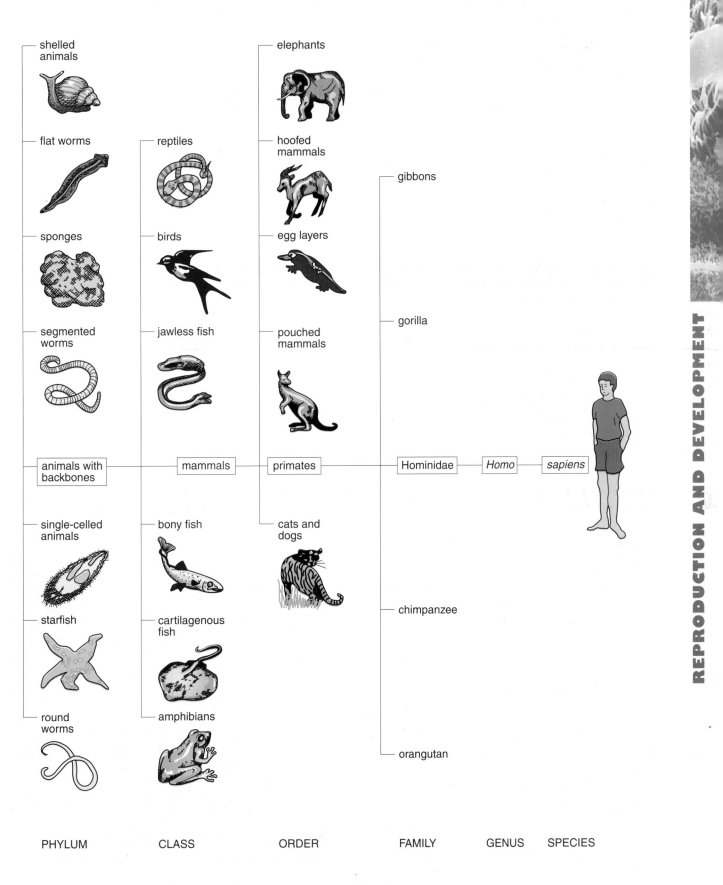

PHYLUM   CLASS   ORDER   FAMILY   GENUS   SPECIES

Figure 2.4.1 The position of humans in the animal kingdom

to produce fertile offspring. In other words, their offspring must be able to breed between themselves.

There may be a great deal of variation within species, e.g. Asians, Africans, Scandinavians, American Indians. However, they all belong to the same species because they can interbreed successfully and produce fertile offspring.

In humans, there are hundreds of features which illustrate variation. Some can be measured and show a continuous trend from one extreme to another. An example of this is **continuous variation** in height in a population. Another type of variation shows two extremes without any intermediate types. Such a case is seen in the ability to roll one's tongue, as shown in the photograph (right).

This is **discontinuous variation**. You can either roll your tongue completely or you cannot. There is no such thing as a 'half tongue-roller'. Tongue-rolling does not change with age or with efforts at learning how to do it.

The girl on the left is a tongue-roller

A great deal of variation can be observed by examining certain characters in students from your school or college. A character is a single feature of an organism which is under observation, e.g. hair colour, eye colour. Variation means the distribution of these characters throughout the group. This can only be seen by collecting all the results for the whole group.

1 Ability to taste phenylthiocarbamide (PTC)

PTC is a chemical which has a bitter taste to some people. It tastes just like the rind of a grapefruit. But you won't know what this tastes like if you are a 'non-taster'!

Prepare slices of fresh grapefruit rind. Touch your tongue with a slice. Record whether you taste anything. Calculate from the class results the percentage of 'tasters' and 'non-tasters'. Is this a clear-cut test? Are there any who cannot say

definitely whether they are tasters or non-tasters? Perhaps your class is not typical. Suggest a reason for this.

Below are data for the frequency of non-tasters in various populations:

Race	Number tested	Non-tasters
Hindus	489	33.7
Danish	251	32.7
English	441	31.5
Spanish	203	25.6
Portuguese	454	24.0
Malays	237	16.0
Japanese	295	7.1
Lapps	140	6.4
West Africans	74	2.7
Chinese	50	2.0
S. American Indians	163	1.2

continued

2 Tongue-rolling

Try to roll your tongue from side to side as shown in the photograph opposite. Find the percentage of tongue-rollers and non-tongue-rollers in your class.

This example is similar to the PTC example. A histogram may be plotted which should resemble the one shown in Figure 2.4.2.

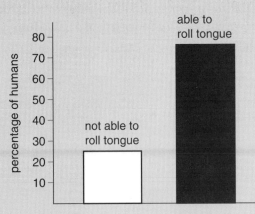

Figure 2.4.2 A possible histogram of results

This illustrates discontinuous variation.

3 Eye colour

Work in pairs. Each member of a pair looks at his/her partners' eyes in good light. Record blue/ brown if they are definitely one of these colours. If not, describe the colour as precisely as you can. Collect the data for blue and brown eyes for the class. You must now decide on a description of the other colours. When you have done this you must decide whether certain colours may be grouped together or whether there is a continuous gradation which does not allow grouping. Is there anyone in the class with two eyes of different colour?

4 Hair

Texture: Classify your partner's hair as being straight or curly and find the percentages of each for the class. Is this a distinct grouping or is there a gradation?

Colour: Hair colour is not at all easy to define. Some colours are black, dark brown, medium brown, light brown, blond, and red. Look at your class' hair colour and modify your colour code if necessary. Record the numbers in each group. Are the differences within groups as great as those between groups? If so, what does this tell you about the variation of hair colour compared with the variation in tongue-rolling ability?

5 Variation in behaviour

Reflexes: One partner holds a 30 cm rule vertically while the other places a thumb and forefinger at each side at the bottom end without touching it. The first partner lets the rule drop without warning, and the second catches it as quickly as possible by closing the thumb and forefinger together. Record to the nearest cm how many cms the ruler fell by measuring the distance from the end of the ruler to the middle of the catcher's thumb and forefinger. Do this test three times without practice and calculate the average length for each member of the class. The more rapid the reflex action, the shorter will be the distance through which the ruler is allowed to fall.

Display the class' results as follows, and construct a histogram by plotting numbers of individuals against classes of distances fallen.

Distance fallen in cms	1	2	3	4	5	6	7	8	9	10
Number of individuals										

6 Height

Record the height of each member of the class, complete a table similar to the one shown below and construct a histogram as you did for reflex times.

Heights cm	150–5	156–160	161–5, etc
Number in each class			

Variations – their causes and significance

Continuous and discontinuous variation

We can use discontinuous variations to sort individuals into distinct groups with no overlapping. In continuous variation there are no distinct groups and individuals are evenly spread over a range of measurements. By plotting histograms in the way suggested in the investigations, we obtain an idea of the limits within which most individuals fall. Also, we can see how the more extreme cases are spread on each side. Before making measurements we should follow these rules:

1   Make sure that the character you are measuring does not depend on another factor. For example, if you are measuring height in a population of humans or length of leaves, the sample must all be of the same age group. The age of an individual will certainly affect these measurements.

2   When measurements involve judgments of degrees of a character, e.g. hair colour, everyone involved in measuring must agree with the judgement and use the same descriptions.

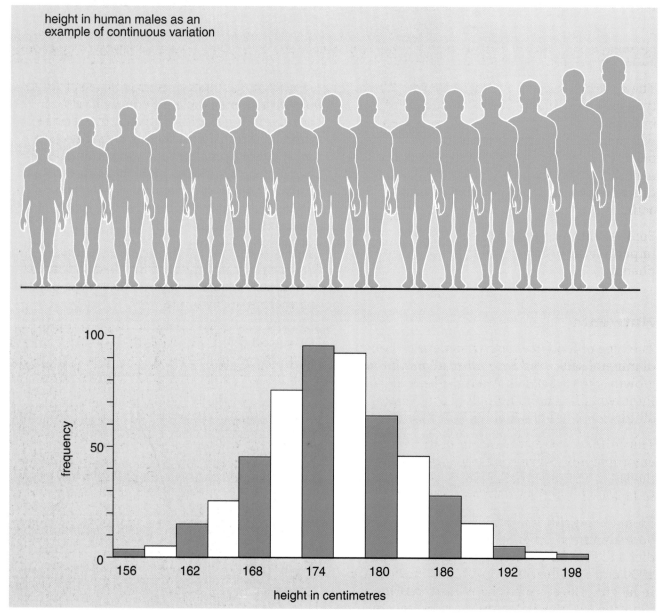

height in human males as an example of continuous variation

Figure 2.4.3 Continuous variation in height

Causes of variation

The outward appearance of an organism depends on a combination of inherited characters and those due to the effects of the environment. The name given to the outward appearance of an organism is phenotype. We can say:

> phenotype = effects of genes +
> effects of the environment

A simple example can illustrate this: imagine a litter of Alsatian puppies. They all have **genes (units of inheritance)** from the same parents. If we compare the growth of these puppies, we might see that one out of the litter becomes less strong and not so large as the rest. The reason could be that it has not been given the correct diet and exercise. In other words, the effects of the environment have influenced its phenotype.

By altering the environment we can alter the phenotype.

On the other hand, a litter of corgis will grow up to be small dogs no matter how well fed they are. The genes determine the phenotype if the environment is kept the same. We cannot alter the genes by changing the environment.

Environmental factors

The environment of an organism means the type of place where it lives. However, the biological meaning contains every influence which is acting on the organism from the outside. It therefore includes food and any other materials taken into the organism, temperature, degree of acidity, light and any physical forces acting on the organism from the outside.

Some examples will serve to show how important these factors are in shaping the phenotype.

Nutrients

The type of food eaten by animals and the types of mineral salts used by plants influence their growth. Figure 2.4.4 shows the effects of improved diet on the growth of some Japanese children.

The children were brought up in different environments. Some were born and raised in the USA. The other data refer to children raised in Japan with and without an improved diet.

From the graphs you can see that the American-born are the tallest at all ages. The improved diet and standard of living in Japan between 1900 and 1952 has also resulted in an increase in height. The same has happened in Britain. Children are now taller and heavier, age for age, than they were 60 years ago.

An abnormal phenotype can also be shown in humans as a result of a deficiency of iodine. Lack of iodine in drinking water may be the cause of goitre formation. The swellings (see page 11) are caused by an enlarged thyroid gland which makes a hormone containing iodine.

Light

The browning of human skin when exposed to light is an obvious change of phenotype due to an environmental change.

A sun tan is an environmental change

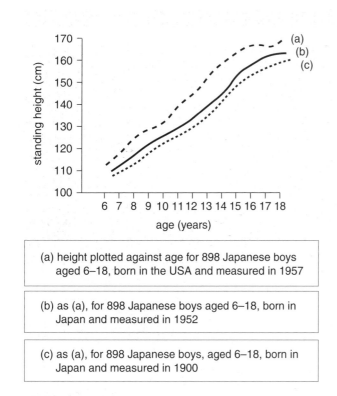

(a) height plotted against age for 898 Japanese boys aged 6–18, born in the USA and measured in 1957

(b) as (a), for 898 Japanese boys aged 6–18, born in Japan and measured in 1952

(c) as (a), for 898 Japanese boys, aged 6–18, born in Japan and measured in 1900

Figure 2.4.4 The effect of improved diet on growth in Japanese children

Vitamin deficiency can also result in abnormal phenotypes, e.g. lack of vitamin D causes rickets (see page 13). The bones become soft and are distorted under the weight of the body.

Chemicals

Some chemicals enter the body from the environment which cannot be called 'food'. They are not needed for growth or energy release and may affect the phenotype. One such chemical, which had a disastrous effect, was thalidomide. This drug was given to pregnant women to prevent morning sickness. Often these women gave birth to deformed children without limbs. The fetus develops in an environment consisting of amniotic fluid and the mother's blood. In these cases, thalidomide exerted its effects by entering the mothers' blood and crossing the placenta.

Temperature

Temperature changes act mainly by affecting the rates of enzyme-controlled reactions in all cells. These effects are shown in the graph (Figure 2.4.5).

The higher the temperature, the faster enzymes work up to a maximum rate. This explains why animals and plants generally grow more slowly at lower temperatures and why most organisms cannot survive at higher temperatures than 45°C.

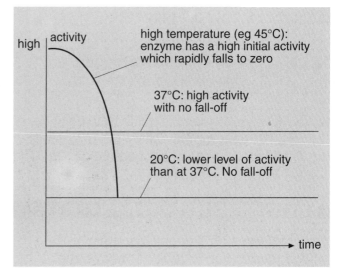

Figure 2.4.5 The effect of temperature on enzyme activity

Physical forces

Training by using muscles more than they are normally used can alter the phenotype. Muscle enlargement is obvious.

Also the effect of training is that the blood vessels to the muscles enlarge to supply more nutrients and oxygen.

This woman has built up her muscles through physical training

REPRODUCTION AND DEVELOPMENT

Natural selection

Charles Darwin

These gulls have found a new food source on a household rubbish tip

With a knowledge of variation, we can now develop the idea of natural selection in the way that Charles Darwin developed it in his famous book *The Origin of Species*, published in 1859. This was written after his famous voyage of discovery on HMS Beagle in 1834. He based his theory on three observed facts and on two deductions from them.

The first fact was stated in 1798 by Thomas Malthus in his *Essays on Population*. Malthus explained that animals and plants have a tendency to multiply at a geometrical rate. This is, in numbers that run 2, 4, 8, 16, 32 and so on. The offspring always tend to be more numerous than parents. We have only to look at the human 'population explosion' to see this in action.

The second fact is that, while all living organisms can increase at a geometrical rate, they seldom do. Few species, apart from humans and some animals and plants dependent on us, have been observed to increase so rapidly for very long. The species

that have done so have often been presented with new opportunities by humans. For example, the gulls feeding on our refuse dumps and rats feeding in our sewers. Another example was seen in the rabbit which was introduced to Australia. There, they had no natural enemies and in a few years became a national nuisance.

From these two facts, Darwin deduced a 'struggle for existence'. More accurately, the competition for a chance to reproduce. Almost everywhere in nature, organisms produce more young than can possibly survive to the age at which they can reproduce. They must compete for food and for all their other needs. For example, we are not completely covered with flies despite their enormous breeding rate.

The third fact is that all living things vary. We have seen this in our studies of genetics. Darwin deduced from these facts a mechanism of **natural selection**. The principle of natural selection states that the competition is for existence between individuals which vary

REPRODUCTION AND DEVELOPMENT

163

Rapid population growth occured when rabbits were introduced to Australia

among themselves. Some individuals must be more likely to succeed than others. Those with favourable variations will be more likely to survive and reproduce than those with unfavourable variations.

A great deal of variation is inherited. Favourable inheritable variations have a better chance of being passed on than unfavourable variations. Natural selection is the principal cause of **evolution**. It is not a physical force, it has no purpose; it is a process that occurs completely by chance, like evolution itself. It is a process that has made, through millions of years, the human brain, the bird's eye and the bat's ear – all from simple cells.

However, although Darwin was convinced that variations within species could be passed on to future generations, he had no knowledge of the causes of variations through mutation.

Variation and mutation

No two living things are exactly alike. If you look casually at a group of organisms of one species, you might think that they were all alike. However, anyone who has to care for animals soon learns to distinguish between them. Furthermore, plants can vary just as much as animals.

A boy tends to resemble his brother or his father more closely than more distant relatives, and he resembles his relatives more closely than his unrelated friends. These resemblances are due mainly to **heredity**. The resemblance of identical twins is due totally to heredity. The differences that enable us to tell one twin from another are due to environment.

Separation of chromosomes in meiosis (see page 138) and the random way in which they recombine also produces variation. Even the simplest of organisms have hundreds of genes and complex organisms have thousands. A tremendous amount of variation is possible from this number of genes. Furthermore, crossing over of parts of chromosomes during meiosis increases the possiblity of variation by making new combinations of genes. The mixing

of chromosomes at fertilisation ensures that all individuals are unique.

Other factors affect development of an organism. A pea plant may inherit tall genes, but will still not grow in poor soil. Regardless of how many times different genes are combined into new arrangements, the results are all variations on an existing pattern. A change in the genes themselves – a **mutation** – will give a new type of pattern.

One of the first mutations studied was one in the fruit fly, *Drosophila melanogaster*. T H Morgan (see page 143) found that in a strain of pure-bred red-eyed flies, there was one with white eyes. This change was caused by a sudden change in one of the many genes controlling eye colour. Since then hundreds of fruit fly mutations have been found and studied.

What is a mutation?

Mutation involves a change in the chemical structure of a gene. Mutations can result from a mistake in gene duplication. For example, in

REPRODUCTION AND DEVELOPMENT

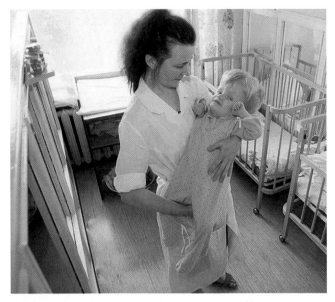

A human gene mutation caused by radiation from Chernobyl

replication, the DNA might pick up the wrong base and so produce a new arrangement. Since genes control the manufacture of proteins, the new gene might make a different protein, or be unable to make a protein at all. This could

result in a change in, or the introduction of, a body characteristic.

Most mutations are recessive and are masked by dominant normal genes. Thus, except for some sex-linked genes (see page 173), mutations do not show up until two of the mutant genes occur in the same individual. (The individual is said to be homozygous for the mutant gene.) Some mutants produce only minor changes in body chemistry. If it is an unfavourable change for the organism it will be lost because the organism with the mutation is unlikely to have offspring which will inherit the mutant gene. Sooner or later, however, the mutation will occur again. An example of a mutation is shown in the photograph (left).

Although we speak of mutant genes and normal genes, the genes we now call normal were once mutants. Because they were favourable, they have been passed on from generation to generation and have become part of the normal collection of genes. Mutation is all part of the process of evolution.

Radiation and genetics

All living things are exposed to a certain amount of radiation. The two sources of this radiation are:

1 Cosmic rays from space.

2 Radioactive materials in the Earth's crust.

This **background radiation** is, in part, responsible for the mutations that occur in all organisms.

Consider radiation as being a series of rapidly-fired mini-bullets. Then think of these bullets hitting molecules that make up organisms. These molecules will be damaged and if DNA is one of them, then the code which is normally inherited can be destroyed or altered.

In fact, radiation is a series of sub-atomic particles (mini-bullets) which leave some atoms with enormous energy. When these particles hit an atom they strip off an electron and release a large amount of energy. So any biological

molecule hit by radiation is destroyed. This, in itself, is enough to cause considerable damage to a living cell.

It has been known for a long time that an increase in radiation causes an increase in the mutation rate. In other words, it causes an increase in the rate at which genes are altered. Indeed, this knowledge has been used in genetic experiments to produce mutations in experimental organisms. This treatment has produced some useful plant varieties.

It is also known that there is no level of radiation that is so low that no change in mutation rate occurs – radiation always has an effect. These findings have real significance for us in our nuclear and technological age. The human race has increased the radiation levels in several ways. First, there has been an increase in the use of **X-rays** for medical purposes. Then there was an increase in the testing of nuclear weapons.

Nuclear test site at Mururoa Atoll, South Pacific

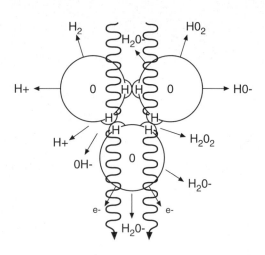

Figure 2.4.6 How radiation causes indirect damage to cells

This brought a whole new meaning to the word 'fallout'. The dumping of radioactive waste from nuclear power stations is a problem which has not been solved with modern technology.

It must be emphasised that this increase in radiation is only a fraction of the naturally occurring background radiation. However, in the same breath it must also be emphasised that *any* increase will increase the mutation rate.

How the damage is done

Radiation may damage a cell by direct action or by indirect action. The result of **direct action** is damage to a molecule e.g. damage to a molecule of DNA by the passage through it of an electron or other atomic particle. **Indirect action** is the change in a large molecule of DNA brought about by highly active pieces of molecules (free radicals or ions). Free radicals are formed by the action of radiation on water molecules.

Figure 2.4.6 shows the various particles that are produced when water molecules absorb radiation energy. Although these free, highly active pieces of molecules are very short lived, they can do a lot of damage to any large molecules that are near them. In this way radiation brings about the mutation of genes.

The different types of radiation

From a biological point of view, the most important damaging radiations are alpha particles, beta particles, and X-rays and gamma rays.

Alpha particles spread their energy very quickly. They form an enormous number of pieces of ions. There is little danger from alpha particles unless they get inside the body. The skin easily stops the particles entering from outside sources. However, if an organism takes in food exposed to alpha radiation, contaminating the body cells, the alpha particles become a serious hazard. The most active tissues in the body are blood-producing bone marrow, the liver, testes and ovaries. Once alpha particles get into these tissues they will do two kinds of harm:

1 The marrow and liver cells will suffer damage – this will shorten the life of the person.

2 The damage done to the sex organs will be inherited via DNA and passed on to future generations.

Products from nuclear reactors are one source of alpha particles and pose a serious problem. For example, plutonium-239 is a long-lasting isotope which can easily enter food chains. Once in the body, it accumulates in the bone and is suspected of causing anaemia and

REPRODUCTION AND DEVELOPMENT

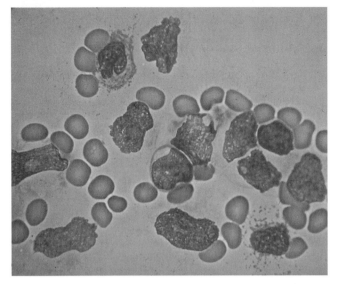

Blood from a person with leukaemia ×795

Testing effluent from Sellafield nuclear power station

leukaemia. Leukaemia is a particularly distressing form of cancer which affects blood cells in the bone marrow. Leukaemia occurs when mechanisms which control blood cell manufacture break down. An imbalance in the correct proportions of red cells, white cells and platelets then occurs.

Nuclear waste leaked from the Sellafield reprocessing plant in Cumbria has been detected in the Irish Sea.

Fish are now being tested for radiation contamination. Grass and seaweed samples near power stations are also analysed to check that the radiation levels are not too high.

Beta particles are high speed electrons and can penetrate the skin. Internally, they are also a hazard. Radioactive fallout contains beta particles as well as alpha particles. Strontium-90 and caesium-137 have attracted considerable attention because they readily accumulate in living organisms. For example, fish caught in the North Sea have been found with caesium-137 in their flesh. Strontium-90 is similar to calcium and becomes localised in bones. Emissions can damage the red bone marrow and cause anaemia and leukaemia. Caesium-137 is absorbed by all cells.

Sampling water for radioactive contamination

Seaweed is regularly checked to make sure radiation levels are safe

REPRODUCTION AND DEVELOPMENT

X-rays and **gamma rays** have great penetrating power. They are a massive hazard when their source is outside the body. Great care must be taken to minimise exposure when X-rays are used. Again, the most active tissues – developing embryos for instance – are most sensitive. X-raying pregnant women is risky for the embryo and is rarely, if ever, carried out.

Nuclear energy, radioactivity and the environment

One of main environmental concerns today is disposal of **nuclear waste** from power stations. The problem of extremely long **half-lives** of radioactive waste chemicals is the main consideration when deciding on disposal. The half-life period is the time that it takes for the radioactive chemical to be reduced by half. It can be thought of as something decaying away over a period of time.

Here is a list of the half-lives of some important radioactive elements. Some of these may enter our environment as a result of human activities:

Table 1 Some radioactive isotopes and their half-lives

Chemical	Half-life
carbon-14	5760 years
phosphorus-32	14.3 days
sulphur-35	87.2 days
calcium-45	165 days
strontium-90	28 years
iodine-131	8.04 days
caesium-137	30 years
plutonium-239	24 400 years

Plutonium's radiation is made up of alpha particles. Note the vast difference between the half-life of this element and the others in the list!

Radiation and you

Can nuclear wastes be buried at sea?

The production of highly radioactive waste is a problem with all technologies which use radioactive materials, including nuclear power stations. This waste must be disposed of safely, in places from where it cannot be recovered. Three main places for the disposal of solid, high-level waste have been suggested. One is the surface of the deep ocean floor, others are

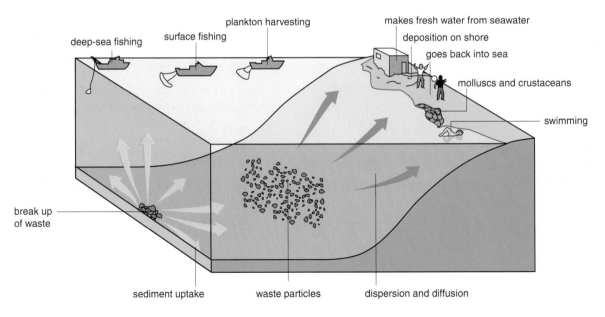

deep-sea fishing · surface fishing · plankton harvesting · makes fresh water from seawater · deposition on shore · goes back into sea · molluscs and crustaceans · swimming · break up of waste · sediment uptake · waste particles · dispersion and diffusion

Figure 2.4.7 Radiation in the marine environment

under the surface of the land and under the surface of the ocean floor. Whichever method or combination of methods is used, there will be routes by which radiation will return to the human environment.

If leakage from containers stored on or under the ocean bed takes place, pollution by radiation will occur and could affect us. Figure 2.4.7 shows the possible routes the radiation could then take as it returns to our environment.

The total radiation that affects the human environment is given in the following table and is shown diagrammatically in Figure 2.4.8.

Source of radiation	Approximate %
naturally from outer space and rocks	87.0
medical uses	11.5
luminous watches, etc.	0.9
weapons testing	0.5
waste dumped at sea	0.1

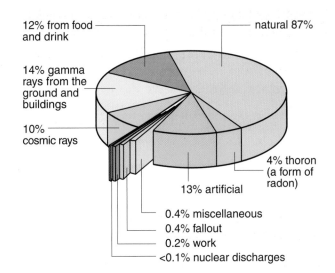

12% from food and drink
natural 87%
14% gamma rays from the ground and buildings
10% cosmic rays
13% artificial
4% thoron (a form of radon)
0.4% miscellaneous
0.4% fallout
0.2% work
<0.1% nuclear discharges

Figure 2.4.8 Sources of radiation

It is not only nuclear power stations that cause problems. Power stations in the UK using fossil fuel produced approximately 0.5 million tonnes of smoke in 1986. This contained varying amounts of carbon dioxide, sulphur dioxide, and oxides of nitrogen. All these pollutants cause problems, not the least of which are acid rain and the greenhouse effect. Available technology can remove 90% of the sulphur dioxide at an approximate cost of 200 million pounds per power station.

Summary

1 Plants and animals show continuous and discontinuous variations which can be measured and expressed graphically.

2 Causes of variation may be genetic or environmental.

3 Charles Darwin is probably the best known naturalist of all time. His major contribution to history was his theory of evolution through natural selection.

4 He based his theory on observations and deductions made during his famous voyage of discovery on HMS Beagle.

5 Variations result from mutations.

6 Radiation can cause mutations which show themselves with disastrous consequences.

7 Our use of nuclear technology may result in accidents, such as that witnessed by the inhabitants of Chernobyl in Russia.

REPRODUCTION AND DEVELOPMENT

Questions for review

1 Explain the difference between continuous and discontinuous variation and give an example of each in humans.

2 What is the meaning of the term 'mutation'?

3 List the major causes of mutation.

4 Why are mutations that occur in body cells not important to the entire species?

Applying principles and concepts

Case study – Chernobyl

In 1986, an accident at the Soviet Union's largest nuclear power station released a cloud of radioactive material high into the atmosphere. The winds then blew this cloud across Poland and Scandinavia. These countries were showered with radioactive chemicals. The direct risks to health were inhalation and skin irradiation but there were other indirect risks. There were restrictions on the consumption of fresh vegetables and milk in several countries.

In March 1989, the following headline appeared in a science magazine:

HEADLESS CALVES BETRAY THE LEGACY OF CHERNOBYL

Here is a summary of the article:

"The effects of the world's worst nuclear accident at Chernobyl are becoming clear at farms just outside the 30-kilometre exclusion zone around Chernobyl.

There has been an alarming increase since 1986 in the number of deformed animals being born on a farm 50 km from Chernobyl. This farm, which has 350 cows and 87 pigs, has records of only three deformed pigs born in the five years before the accident and no deformed calves. In 1987, 37 pigs and 27 calves were born with serious deformities. During the first nine months of 1988, a further 41 deformed pigs and 35 deformed calves were born. Some calves lacked heads, limbs, eyes or ribs. Most of the pigs had deformed skulls.

Radiation at the farm was 148 times as high as background level. Food is still delivered to the area from elsewhere. However, local people still consume home-produced milk, fruit and vegetables. Cattle still eat locally produced fodder.

Concern is now growing about the possible effects on people. For instance, women from the area are advised not to have children. The average annual number of new cancer cases, especially of the lip, has doubled since the accident.

Farmers use pressurised cabins on tractors to protect themselves from radioactive dust from the fields. Medical workers have found the thyroid glands of more than half of the children were affected by radiation."

1 Explain how radioactive materials get into milk and vegetables.

2 Besides the potential hazards of accidents in nuclear power stations, state another problem which arises from generating electricity by nuclear means rather than by using tidal or solar energy.

3 One of the radioactive materials was iodine-131. Explain the link between this and its accumulation in the thyroid glands of people affected by the radiation.

4 Give a genetic explanation for the birth of deformed animals that took place two years after the accident.

REPRODUCTION AND DEVELOPMENT

When things go wrong

Learning Objectives

By the end of this chapter you should be able to:

- Describe some of the major genetic disorders in humans
- Recognise the significance of disorders due to abnormal genes or chromosomes
- Realise that some disorders are a

result of dominant genes and some are a result of recessive genes
- Understand the importance of genetic engineering in the production of hormones

Genetic disorders

You know by now that we are far from understanding all of the muations that interfere with the way the body works. It is possible to classify at least two genetic effects. One group consists of mutations that stop the body making certain essential proteins, e.g. enzymes, hormones, or proteins making up the structure of cells.

The second group is related to the first in that it often involves the lack of certain enzymes. However, the main problem is the accumulation in the body of chemicals which normally would be broken down. An example of this type is the accumulation of a chemical called phenylalanine in the blood. It leads to the condition called **phenylketonuria (PKU)** (see page 175).

Inborn errors

Inheritance of abnormal proteins can be seen in various forms of blood disorders in humans. Perhaps the best known example of a genetically controlled blood defect is **haemophilia**. This is a condition in which the blood fails to clot, or else it clots very slowly. People with extreme cases of haemophilia can bleed to death from normally minor injuries.

The chemistry of blood clotting is very complicated and involves over a dozen different enzyme-controlled reactions. If one or more of the enzymes fails to be made by the body, then blood clotting is hindered or perhaps (in rare cases) prevented altogether.

A royal complaint

Haemophilia is a genetic defect with a royal history. The gene for the problem became so widely distributed in European royalty during the nineteenth and twentieth centuries that the course of history was affected – especially in Spain and Russia. The gene for the condition probably first appeared as a mutation in Queen Victoria, since there is no record of haemophilia in her ancestry. Because of marriages among the royalty of Europe, the gene became distributed in a number of royal families. Figure 2.5.1 shows the pedigree of the distribution of

haemophilia in Queen Victoria's descendants. Note that the present royal family of the United Kingdom is free of the gene.

Men only

Haemophilia only appears in males (except in very rare circumstances), although the condition is inherited from the mother. This

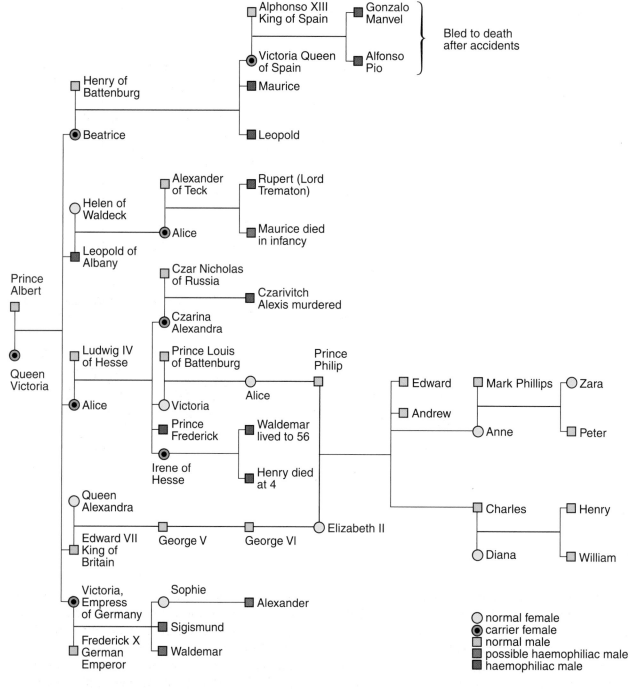

Figure 2.5.1 Haemophilia pedigree

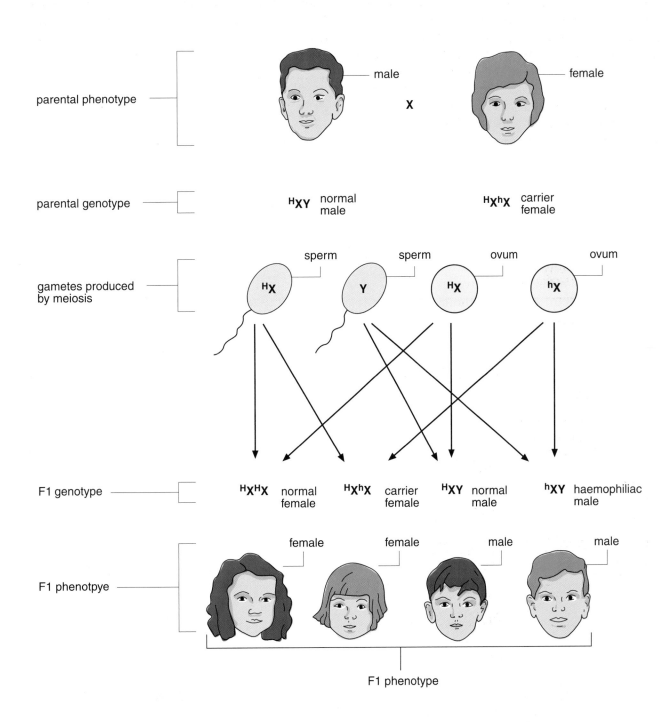

Figure 2.5.2 Inheritance of haemophilia

form of inheritance is called **sex-linked recessive**. This means that the gene is carried on the X sex chromosome. Males have only one X, whereas a female has two X chromosomes. The condition is masked by the normal gene on one of these. Males in addition have one Y chromosome. So, if an affected

male becomes a father, then all his daughters will be carriers and all his sons normal (they receive their X from the normal mother). The condition can only be carried on the X chromosome. Therefore an affected male must be X^hY (h = gene for haemophilia and H = normal blood):

Parents: $X^hY \times X^HX^H$

Children: X^hX^H, X^hX^H, X^HY, X^HY

X^hX^H is a carrier female and X^HY is a normal male.

If a carrier female becomes a mother, then the chances are that half of her sons will be unaffected and half her daughters will be carriers:

Parents: $X^hX^H \times X^HY$

Children: X^hX^H, X^HX^H, X^hY, X^HY

The problem is that, in this situation, there is no perfect method of telling the carrier daughter from the normal daughter. Obviously, it is extremely difficult to give guidance on family planning to such women. The councillor can only say with confidence that daughters of an affected man will be carriers. The daughter of a haemophiliac female carrier cannot be identified as a carrier herself until she has given birth to a haemophiliac son.

In approximately one third of all patients no history of the condition can be detected in the family ancestry. This may be due to the transmission through several generations of female carriers without an affected male being born. Furthermore, all record of a previously affected male could be lost to the memory of that family. Alternatively, the condition may have arisen by a gene mutation.

It is not possible to provide accurate statistics for all countries, but in Europe and in North America the incidence of haemophilia is approximately one per 8 000 to 10 000 of the population. There have been major improvements in treatment of the condition and the chances of survival have undoubtedly improved in recent years. Also, the number of recognised patients is increasing leading to more meaningful councelling.

The common errors

Possibly the most common human genetic disorder is **cystic fibrosis**. It is an inherited condition which affects the pancreas, and the bronchioles of the lungs. It results in failure of the pancreas to function properly, intestinal obstruction, sweat glands and salivary gland malfunction. Cystic fibrosis is one of the most common fatal diseases of childhood and is inherited as a recessive gene:

If C = the gene for a normal pancreas and bronchioles and c = the recessive gene for cystic fibrosis, a person suffers from the disease only if he or she has two genes for cystic fibrosis, i.e. the genotype cc. A person with the genotype Cc is a carrier of the disorder but does not suffer from the disease. There is therefore a one in four chance of a child suffering from the complaint if two carriers have a child:

carrier man Cc $\times$ carrier woman Cc

gametes	C	c
C	CC	Cc
c	Cc	cc

The condition occurs once in about 2000 births. It accounts for between 1% and 2% of admissions to children's hospitals. Many children die as a result of pneumonia, though some survive to adulthood.

A blighted gene

Neurofibromatosis – NF for short – is one of the commonest inherited diseases. It affects about 18 000 people in the United Kingdom. About 200 babies are born with the condition each year. It varies in severity from just a few coffee-coloured skin spots to life-threatening tumours. Indeed many people do not even realise they have it.

NF is caused by a mutant gene on the human chromosome number 17 (see Figure 2.3.9). How this causes the many features of the disease is not known. The most common sign is the growth of small nodules on the skin which tend to develop at puberty.

The first symptoms are coffee-coloured patches on the skin. These are followed by brown spots on the iris of the eye. At the beginning of puberty, nodes develop. They increase in number throughout life but they rarely cause major health problems. It is the complications of the disease which are more serious. These include short stature, curved spine and learning difficulties. In 1% of cases victims suffer from large facial tumours. There is a 2–3% chance of nodules becoming cancerous.

A missing enzyme

There is an enzyme which certain people cannot make. This enzyme is responsible for changing a chemical called **phenylalanine** to an amino acid used by the body. Phenylalanine occurs in meat, fish, cheese, eggs, wheat and butter. In the absence of the enzyme which acts on it, phenylalanine accumulates in the tissues of the body. Some is converted into a dangerous chemical which causes damage to the central nervous system. It leads to a person being mentally retarded and is due to a single mutant gene. About one case in 15 000 babies suffers from the problem, called **phenylketonuria**, and synthetic substitutes for protein must be given in the diet.

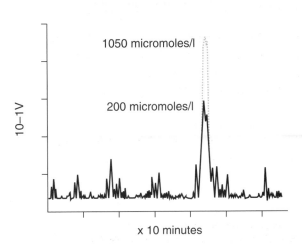

Figure 2.5.3 Abnormal levels of phenylalanine in babies
Normal blood sample (black bar)
Abnormal blood sample (grey bar)

A dominant problem

In 1872, an American doctor, Fraser Roberts, wrote the following observation in his note book: "The boy George Huntington, driving though a wooded lane in Long Island while accompanying his father on professional rounds suddenly saw two women, mother and daughter, both tall and very thin, both bowing, twisting and grimacing." Fifty years later George Huntington made a major contribution to science by describing and studying the inherited disease, now known as **Huntington's chorea**.

It is an inherited disease characterised by involuntary muscular movement and mental deterioration. The age of onset is about 35 years. The majority of those affected can therefore produce a family before being aware of their own condition. It is transmitted by a dominant gene and both sexes can be equally affected. An estimate of its frequency is roughly five in 100 000 and affected people are heterozygous.

If HC = gene for Huntington's chorea and hc = gene for normal:

mother HChc × father hchc

gametes	hc	hc
HC	HChc	HChc
hc	hchc	hchc

Theoretically, two parents could be affected and a homozygous child could be produced. The effect of this would probably be lethal. The gene is so rare that it is most unlikely to occur in the homozygous condition.

A gene disorder – sickle-cell anaemia

This is an often fatal condition which is quite common in West Africans. If there is a low level of oxygen in the blood, red blood cells of a person suffering from this disorder collapse into a sickle shape and may form blockages in blood vessels.

The disease is inherited as a single mutant recessive gene. If a child inherits the gene from both parents it has only a 20% chance of surviving into adulthood. The gene that controls the formation of haemoglobin is not formed properly. Therefore the haemoglobin in the sickle-shaped cells is not very good at carrying oxygen.

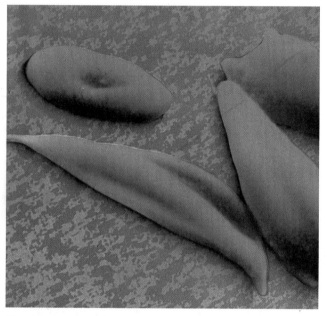

Blood from a person with sickle cell anaemia × 19924

If this is so important you might think that, over thousands of years, the mutant gene would have disappeared from human chromosomes because carriers would die before having children. However, under certain circumstances, it is an advantage to have some sickle-shaped red blood cells. The reason is because the malarial parasite is less likely to attack people who are heterozygous for the mutant gene.

For example, let us assume that the gene for haemoglobin is Hb and that the gene for sickle-cell haemoglobin is HbS. Then a normal person will have the haemoglobin genotype HbHb. A person with the sickle-cell disorder will have the genotype HbSHbS and usually dies. However, a person with the genotype HbSHb can still survive and will be resistant to malaria.

If two people, each with the genotype HbSHb have a child, then the predicted genotype of the child can be shown as follows:

Parents: mother HbSHb × father HbSHb

gametes	HbS	Hb
HbS	HbSHbS	HbSHb
Hb	HbHbS	HbHb

It can be seen that there is a one in four chance of the child dying with the sickle-cell disorder (HbSHbS). There is a one in two chance of the child being a carrier of the sickle-cell gene but probably surviving (HbSHb). There is a one in four chance of the child being perfectly normal for the haemoglobin gene (HbHb).

An extra chromosome

Down's syndrome is caused by the presence of an extra twenty-first chromosome in all body cells (Figure 2.5.5).

The extra chromosome results from failure of the chromosomes (pair 23) to separate properly at meiosis, probably during egg formation. The disorder results in some mental retardation and some abnormal physical features such as heart structure, enlarged tongue, and weak muscles. About one in 600 babies are born with Down's syndrome, but this will vary according to the age of the mother. With mothers under 35, less than one baby in 1 000 has the syndrome, but at 45 and over, there is a chance of one in 60.

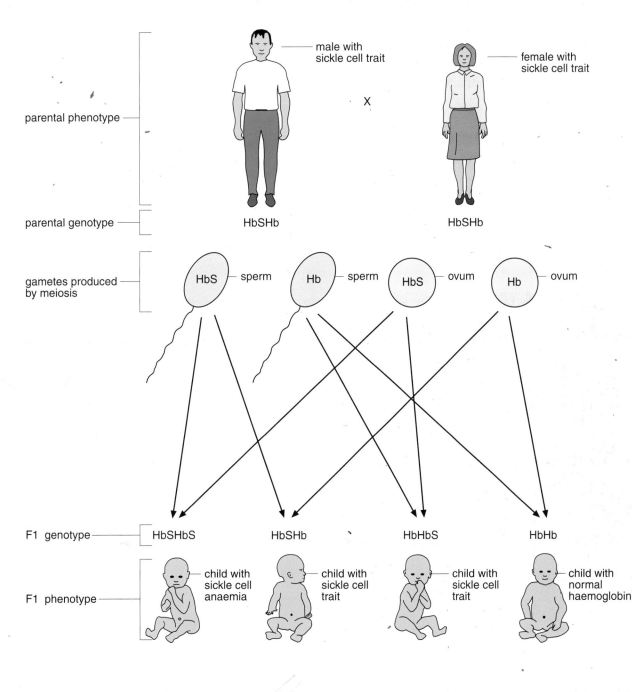

Figure 2.5.4 Inheritance of sickle cell anaemia

Genetic engineering provides hope

The principle of genetic engineering relies on isolating a gene from one organism and putting it into another of a different species. But why should you want to do that in the first place?

The answer is because the gene might be responsible for carrying out a very useful function. For example, scientists often isolate genes from human chromosomes which control the production of hormones. They put these useful genes into bacteria or yeasts.

The idea of transferring the human genes to the bacteria is to increase production of the useful

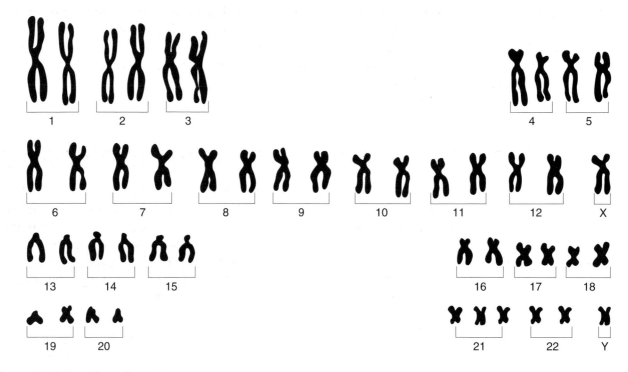

Figure 2.5.5 Down's syndrome karyotype

REPRODUCTION AND DEVELOPMENT

substance. The microbes multiply very rapidly and can be cultured relatively cheaply. In fact they can provide almost unlimited amounts of substances that are practically unobtainable in bulk in any other way.

The process starts with biological 'scissors' called **restriction enzymes**. These cut chunks of DNA at points on chromosomes where useful genes are known to exist. The restriction enzymes therefore enable the scientist to cut out very precisely the gene that is needed. It may be one out of hundreds on a particular chromosome. The next stage is to put the gene into a bacterium.

It is not put directly into the bacterial chromosome. Instead, genetic engineers use a circle of DNA called a plasmid. These are normally present in bacteria but are largely independent of the rest of the cell. Plasmids, like chromosomes, carry bacterial genes which control the microbe's metabolism. The plasmid is cut open with restriction enzymes and the foreign gene inserted. The break is sealed with another enzyme called a **ligase** (an enzyme which binds chemicals together). This process makes a mixed molecule called recombinant DNA.

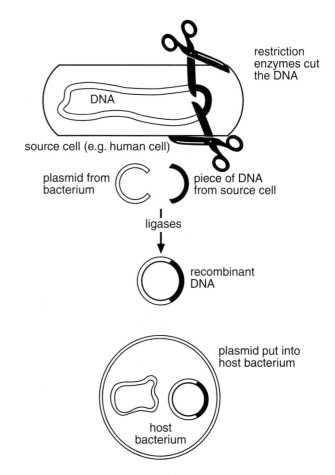

Figure 2.5.6 Recombinant DNA technology

The altered (infective) plasmids are then mixed in a test tube with bacteria which do not have any plasmids. Some plasmids move inside these bacteria. The infective plasmids carry the foreign gene inside the cell where it can instruct it to make the required protein (Figure 2.5.6).

The process is regarded as a success only when the gene is expressed. This means that the host cell obeys the instructions carried by the foreign gene and makes human proteins.

Life-saving genetics

There is a hormone called **insulin** which is made by certain cells in the pancreas. Its function is to keep the concentration of **glucose** in the blood at a constant level (0.1 g per 100 cm^3 of blood). If the glucose level falls much below this, the body does not have enough fuel to function properly. A glucose level above the normal concentration also disturbs body functions. In particular, the kidneys fail to cope and glucose is lost in the urine. The body is eventually drained of fuel.

Insulin plays a vital part in getting rid of excess glucose from the blood. It helps to do this in a number of ways but one way is to change glucose into an insoluble form of carbohydrate called **glycogen**. This is then stored in the liver until a new supply of sugar is needed. Without enough insulin people cannot control their glucose level and suffer from **diabetes**. Fortunately, it is possible to control this condition by injecting insulin into a diabetic person.

Until recently the insulin used for this treatment was taken from the pancreas of various mammals. It was difficult to produce enough insulin in this way. Genetic engineers have solved the problem of mass production by using bacteria. The method is summarised in Figure 2.5.7.

Figure 2.5.7 Insulin production

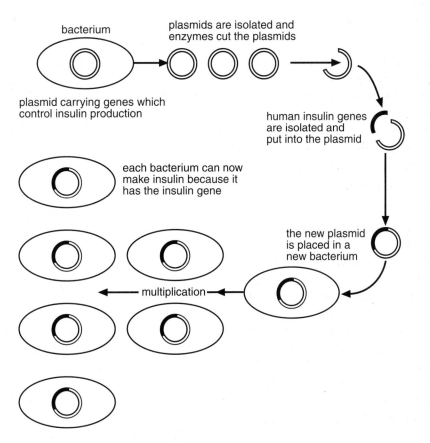

bacterium

plasmids are isolated and enzymes cut the plasmids

plasmid carrying genes which control insulin production

human insulin genes are isolated and put into the plasmid

each bacterium can now make insulin because it has the insulin gene

the new plasmid is placed in a new bacterium

multiplication

REPRODUCTION AND DEVELOPMENT

Summary

1 Occasionally, mutations lead to inheritable abnormalities in humans.

2 Such genetic disorders may be caused by altered genes or altered chromosomes.

3 Common errors in chromosomes can show as symptoms which often require medical treatment.

4 In genetic engineering, useful genes, responsible for production of key chemicals in our bodies, can be taken from one organism and put into another.

5 These organisms are usually bacteria because such organisms can multiply rapidly and produce vast quantities of useful chemicals for us.

Questions for review

1 List some diseases that are associated with genes.

2 What is sickle-cell anaemia?

3 What chromosome abnormality is the cause of Down's syndrome?

4 Why is it so difficult to prevent the gene for Huntington's chorea from being inherited?

Applying principles and concepts

1 A woman has cataracts in her eyes. This is caused by a dominant allele and she is homozygous. The father is normal and homozygous. What will be the predicted phenotypes and genotypes of their children?

2 A pedigree for far-sightedness is given below. Shaded symbols indicate the presence of the disorder.

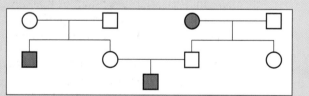

Figure 2.5.8

a) Is the disorder dominant or recessive? Explain how you can tell.
b) Give the genotypes of as many of the individuals in the pedigree as you can.

3 The pedigree shown indicates the occurrence of phenylketonuria (see page 171). Shaded symbols indicate the presence of the disorder.

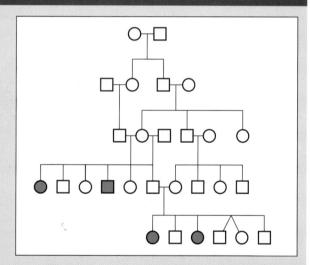

Figure 2.5.9

a) What is the evidence to show that the disorder is recessive?
b) Note the cousin marriages. Why might cousin marriages produce an increased risk?

4 Haemophilia is controlled by a recessive sex-linked gene. Could both a father and son be haemophiliac? Explain your answer.

Sample examination questions

I a) Copy the diagram below and label a part which
   i) produces hormones
   ii) widens during birth
   iii) contracts during birth.

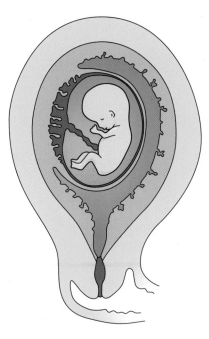

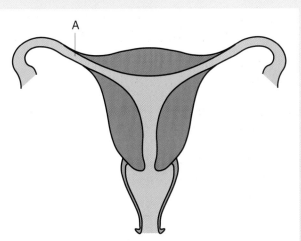

The graph shows how the uterus lining varies in thickness with time.
Fertilisation took place on the 16th day of the second menstrual cycle.

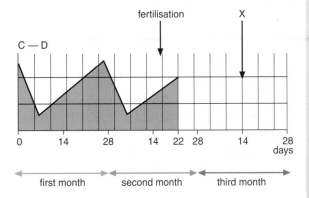

b) Read through the following passage and answer the questions that follow:

The amniocentesis test can be carried out on a pregnant woman. A surgeon inserts a long hollow needle into the woman's uterus (womb) and withdraws some of the amniotic fluid, for tests.
   i) What precaution must the surgeon take before piercing the skin?
   ii) What part of the cells from the amniotic fluid is examined?
   iii) Name one condition (or disease) that can be detected by this test.
   iv) State one risk associated with this test.
   v) State the function of the amniotic fluid.
WJEC, 1992

2 The diagram represents part of the female human reproductive system.

a) Copy and complete the diagram by drawing the ovaries.

b) i) Name structure A.
   ii) Mark with an X the region where sperm normally fertilise egg cells.

c) State the process which took place between C and D.

d) Copy and complete the graph to point X to show what happens to the lining of the uterus after day 22 in the second month.

e) Explain why it is important that the uterus lining changes in the way shown.
WJEC, 1994

3 a) In 1970, 1978 and 1986 surveys were arranged to ask people about their usual method of contraception. The results are shown overleaf.

   i) Which was the most common method of contraception in 1978 and what percentage of people used it as their main method of contraception?
   ii) What percentage of people surveyed in 1986 used the condom as their main method of contraception?

REPRODUCTION AND DEVELOPMENT

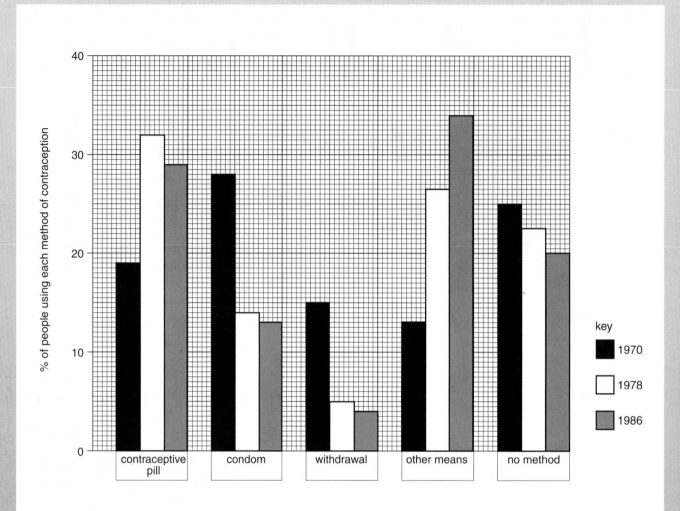

iii) Briefly explain the biological principles involved in the use of the condom and the contraceptive pill as methods or preventing fertilisation.

iv) If the survey had been repeated in 1994 it would have been expected to find that the condom had become more widely used as a method of contraception. Suggest a health-related reason for this.

b) The diagram (right) shows the technique of amniocentesis on a pregnant woman. A needle is inserted, using a local anaesthetic, through the woman's abdomen and a sample of the fluid from around the area of the tip of the needle is withdrawn. This fluid contains cells from the baby, which can then be examined.

What is the name and the normal function of the fluid which has been withdrawn?

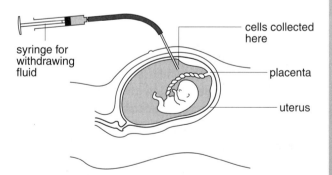

c) The baby's cells are obtained by amniocentesis are prepared so their chromosomes can be studied. The figure opposite shows a drawing of the chromosomes from one of these cells.

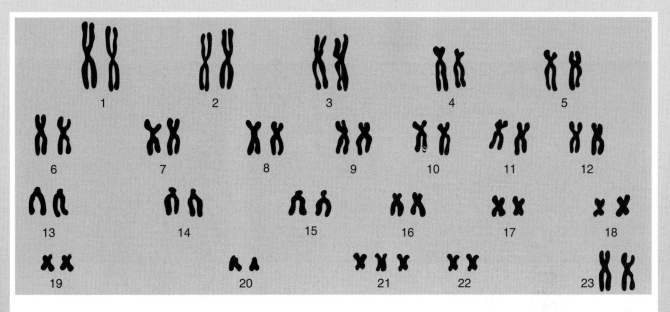

i) What is the sex of this baby? Explain how you can deduce this.

ii) By studying these chromosomes, the baby can be seen to have inherited a genetic disease. What abnormality is shown in the drawing? Suggest the name of the disease that it may cause.

SEG sample paper, 1995/96

4 The graph represents the rate of height increase for boys from one to seventeen years of age.

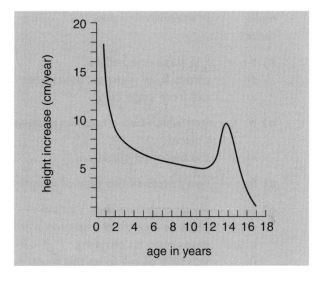

a) Describe the rate of height increase between years one to ten.

b) What name is commonly given to the period represented by the small peak between thirteen and fifteen years?

c) State three other changes that are taking place in the male body during this period.

d) State two changes you would expect to be taking place in the female body from the age of ten to thirteen years.

WJEC, 1994

5 a) Albinos have no pigment in their skin and so look very pale. This condition is caused by the recessive allele, a. Normal skin colour is due to the presence of the dominant allele, A. The pedigree and the key overleaf show the inheritance of this condition through three generations of a family.

i) What is the genotype of:
Jill's mother?
Jack?

ii) Explain how you were able to determine Jack's genotype.

iii) Explain how Lisa and Ruth can have the same phenotype but have different genotypes.

iv) Jack and Jill are planning to have a fourth child. Draw a Punnett square and use it to find the probability of this child being an albino.

b) Red/green colour blindness is another human characteristic which is inherited through recessive alleles. Unlike albinism, red/green colour blindness is a sex-linked characteristic.

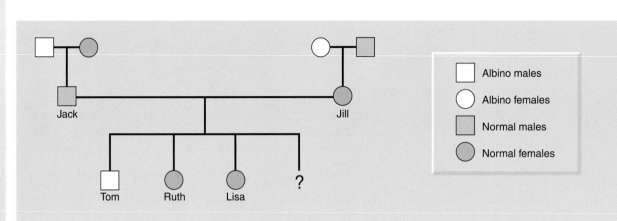

i) Why is red/green colour blindness described as a sex-linked characteristic?

ii) Explain why this colour blindness is much more common in males than in females.

iii) Chris, John and Patrick are triplet boys. John and Patrick are identical twins and have normal vision, but Chris suffers from red/green colour blindness.
  1 Explain how non-identical twins occur.
  2 Explain fully the genetic basis for the difference in the boys' colour vision.

c) The table below shows the percentage risk of a twin developing a disease if the other twin suffers from it.

	Percentage risk to the other twin	
Disease	**Identical**	**Non-identical**
diabetes	50	10
epilepsy	81	21
influenza	27	26
salmonella	47	45
schizophrenia	79	15
tuberculosis	29	12
ulcers	45	23

Use the information in the table to answer the following questions:

i) Name two diseases which do not appear to be genetically inherited.

ii) Name two diseases which are most likely to be genetically inherited (and explain why you chose these diseases).

NICCEA, 1993

6 The diagram opposite shows red bone marrow tissue producing mutant cells which multiply in an uncontrolled way and become cancer cells.

a) If the cell in stage one had 46 chromosomes, how many chromosomes are in a cell from stage two?

b) i) Between which two stages has mutation taken place?
   ii) What is meant by mutation?

c) Suggest two causes of this type of mutation.

d) Sometimes an organism shows a new characteristic as a result of mutation which it then passes onto its offspring.
   i) In which type of cell must this mutation have occurred?
   ii) Explain how natural selection may cause the mutation common several generations later.

WJEC, 1994

REPRODUCTION AND DEVELOPMENT

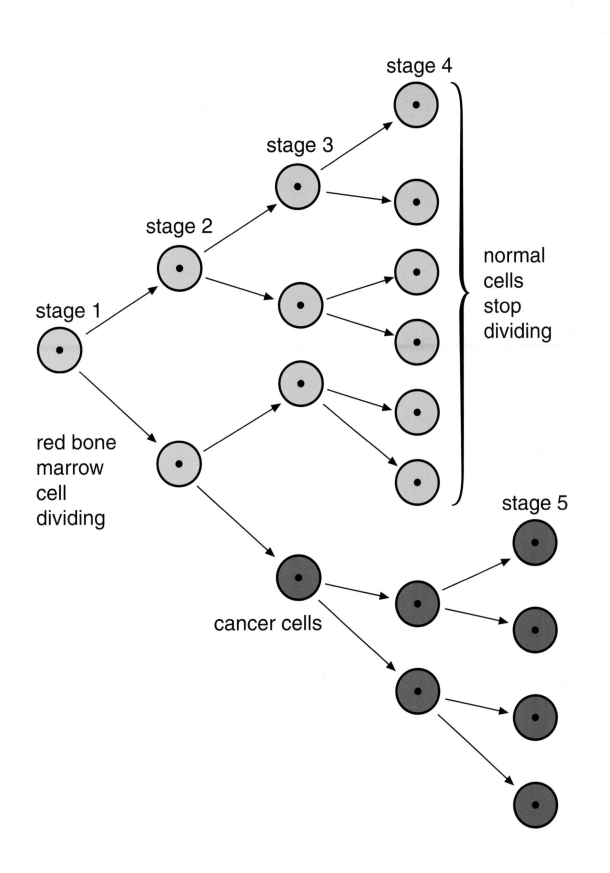

stage 4

stage 3

stage 2

stage 1

normal
cells
stop
dividing

red bone
marrow
cell
dividing

stage 5

cancer cells

HUMANS AND THEIR ENVIRONMENT

The relationships between humans, other organisms and their environment

Learning Objectives

By the end of this chapter you should be able to:

- Understand why photosynthesis is imporatnt to humans
- Demonstrate factors which limit photosynthesis
- Understand the significance of food chains, food webs and food pyramids
- Realise that pyramids of energy are more useful in understanding energy

flow than pyramids of numbers or pyramids of biomass
- Have an awareness of the extent of human impact on the environment including the effects of populaton growth and pollution of air, water and land

Dependence of humans on other organisms

Humans, like all other mammals, depend on plants either directly or indirectly for food. This is because plants can change the energy from sunlight to chemical energy in food. **Photosynthesis** is a process in which carbon dioxide and water, in the presence of light and chlorophyll, are changed so they become the basic building blocks of energy-containing foods. The energy from sunlight is changed to chemical energy in food. Photosynthesis occurs only in cells which contain the green pigment, **chlorophyll**. We can call these cells 'self-fueling'.

Only the self-fueling cells, through photosynthesis, can make **glucose**. With very few exceptions, all other organisms depend on these cells for energy. The dependence may be direct, as in the diet adopted by strict

vegetarians and vegans, or it may be indirect when a person eats beef or drinks milk from a cow. In the end, all human life depends on photosynthesis.

To understand this dependence more fully, imagine that photosynthesis suddenly stopped. Suppose, for example, that Earth was somehow cut off from sunlight. No glucose could be formed and green plants would soon die. Other organisms might survive for a time. Carnivores could eat herbivores, and fungi and bacteria could eat dead organisms but, eventually, food would run out and almost all life would end.

What if the sun kept on shining, but all the green plants were somehow removed? The result would be the same. The remaining organisms would have light energy all around

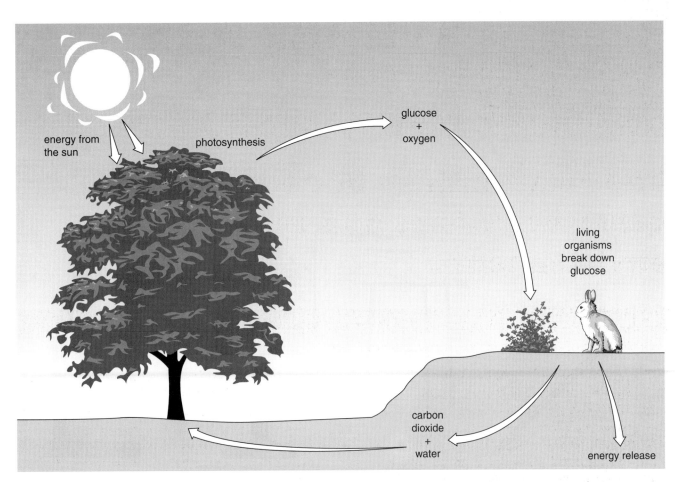

Figure 3.1.1 Energy and living things

them but they could not use it. Eventually all would die.

In the living world, matter is recycled constantly. Simple molecules are organised into complex ones, and the complex substances are broken down to simple ones. This process has been repeated over millions of years, but it requires energy at its source. The energy is sunlight (Figure 3.1.1).

The word, photosynthesis, defines itself. *Photo* refers to light, while *synthesis* means the building up of complex substances from simple ones. The simple substances are carbon dioxide and water. The complex substance is glucose, at first, but this can be the basis of further building into all classes of food except mineral salts. In addition to carbon, oxygen and hydrogen, supplied by carbon dioxide and water, proteins always require nitrogen from the nitrogen cycle (see page 214). Oxygen is given off as a waste

product of photosynthesis and is used for **respiration** in plants and animals.

(in the presence of chlorophyll)				
$6CO_2$	+ $6H_2O$ + light energy	$\rightarrow$	$C_6H_{12}O_6$	+ $6O_2$
carbon dioxide	+ water + light energy	$\rightarrow$	glucose	+ oxygen

From this equation we can see that carbon dioxide, light and chlorophyll are limiting factors. This means that an increase in the quantity of any one of these factors will cause an increase in the rate of reaction. Or, put another way, a decrease in the quantity will limit or decrease the rate of reaction.

It is relatively easy to show that light or carbon dioxide are limiting because, by using a common water plant, we can measure the rate of photosynthesis. We can do this by counting the bubbles of oxygen-rich air given off by the plant when it is supplied with varying light intensity and carbon dioxide concentration.

1 *Plan investigations into the effect on the rate of photosynthesis of:*
a) light intensity
b) carbon dioxide concentration.

For this investigation, use the following apparatus together with a metre rule and a stop watch.

Make a written plan and remember

that all factors influencing the rate of bubble production, except the one you are investigating, must remain constant. Record your observations in a way that will illustrate any relationship between light intensity and the rate of bubble production. Evaluate your method and make a conclusion from your observations.

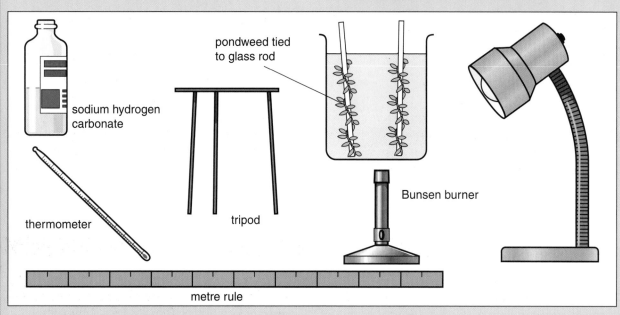

Figure 3.1.2 Experimental set-up

For the second investigation use similar apparatus but include a number of beakers each containing Canadian pond weed (Elodea). You can vary the concentration of carbon dioxide in the beakers by adding different concentrations of sodium hydrogen carbonate.

2 *To demonstrate that the bubbles produced during photosynthesis are rich in oxygen use the following principle. Rusting of metals only occurs in the presence of oxygen. Providing that there is an oxygen-free environment at the start, you can apply this principle to demonstrate that oxygen is produced as a result of photosynthesis.*

Plan this demonstration taking care to use a control tube. After an hour,

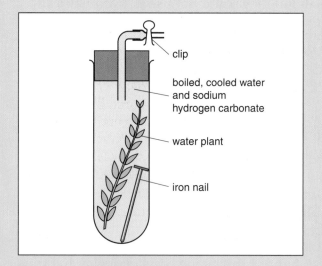

Figure 3.1.3 Apparatus

observe the nails in the tubes. Evaluate your method and make a conclusion based on your results.

3 To demonstrate that starch is a product of photosynthesis a plant must be kept in the dark for two days so that all the starch originally present will be converted to sugar and passed out of the leaves to the stem and roots. This is called destarching a plant.

After such a destarched plant is placed in bright light for several hours, it can be used to demonstrate that starch is a product of photosynthesis, as follows:
a) Remove a leaf from a plant.
b) Hold the leaf with a pair of forceps and put it into boiling water to make burst the chlorophyll-containing chloroplasts and the cell membranes.
c) To remove all the chlorophyll, soak the leaf in boiling ethanol. Use a beaker of boiling water as a water bath to heat a tube of ethanol. Ethanol must never be heated with a direct flame because it is highly flammable.
d) When the leaf is colourless, after removal of the chlorophyll, wash it in cold water to allow it to be permeable to iodine (a test for starch) (see page 28).
e) Add iodine to the leaf in a Petri dish.
f) Observe and record the results.

4 To demonstrate that chlorophyll is necessary for photosynthesis, use the principle of the previous demonstration but use a variegated leaf. This is a leaf which is partly green and partly white. What do you conclude from your results?

Food chains

Plants are the first links in all food chains because they change energy in sunlight into stored chemical energy. When plants are eaten by herbivores some of the energy is passed to this next link in the food chain. When the herbivore is eaten by a carnivore, the process of energy transfer is repeated. Energy passes in this way from carnivores to scavengers, and to decomposers which feed on dead organisms. However, not all the energy stored by a herbivore is stored by the carnivore that feeds on it. Much is used in life processes such as movement, growth and reproduction. Some is also wasted as heat during respiration. Only the left-over energy becomes stored.

When we eat a fish such as tuna we can consider the food chain through which energy flows. Energy from the sun is first used by plant plankton (microscopic algae). It then passes to animal plankton, then to small fish, then to larger fish, then to tuna, then to us. There is no predator to eat us so we are the top carnivores in this food chain.

plant → animal → small → large fish → tuna → humans
plankton plankton fish

In fact you could list hundreds of other species which could be food for the animals mentioned above and draw arrows to show the energy flow. Your diagram would look more like a web than a chain. For this reason, inter-linked food chains are called **food webs**.

Food pyramids

Feeding relationships can be illustrated as pyramids. These can tell us more about the energy that is available to organisms living in a measured area or volume. There are three ways to represent these pyramids:

1 A **pyramid of numbers** which shows the number of organisms per unit area or volume in each feeding layer.

2 A **pyramid of biomass** which shows the dry

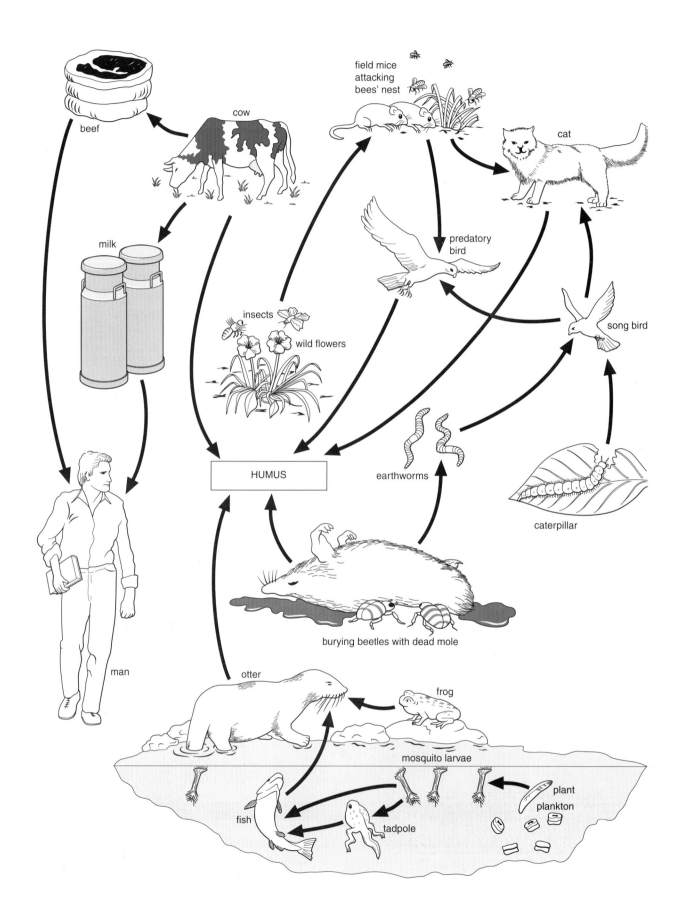

Figure 3.1.4 A food web

mass of organic material per unit area or volume at each feeding level.

3 A **pyramid of energy** which shows the flow of energy through the feeding levels.

The first two methods can be misleading as true pictures of energy flow for the following reasons:

Pyramids of numbers

Note that one is for a field of grass and the other is for a woodland (Figure 3.1.6). The shapes are totally different even though they are made by counting individual organisms. In the woodland, each tree can support thousands of animals and so the base of the pyramid is smaller than the next level.

Pyramids of biomass

Note that the one with plant plankton as a base has a larger first consumer level than base. This is because the biomass of plankton can change seasonally. No account of time has been taken.

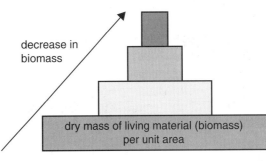

decrease in biomass

dry mass of living material (biomass) per unit area

A pyramid of biomass for a woodland community

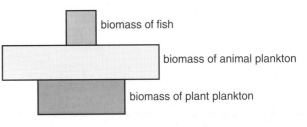

biomass of fish

biomass of animal plankton

biomass of plant plankton

The seasonal change in plankton biomass can result in a pyramid with an unusual shape

Figure 3.1.5 A pyramid of biomass

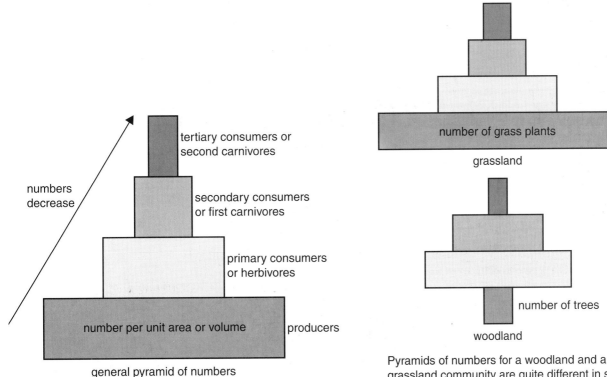

numbers decrease

tertiary consumers or second carnivores

secondary consumers or first carnivores

primary consumers or herbivores

number per unit area or volume   producers

general pyramid of numbers

number of grass plants

grassland

number of trees

woodland

Pyramids of numbers for a woodland and a grassland community are quite different in shape

Figure 3.1.6 A pyramid of numbers

The number of links in a food chain is limited by the chemical energy available. As the amount of chemical energy at each feeding level becomes smaller, so does the amount of living material that can be supported by that level. When the chemical energy dwindles to nothing, the food chain (and the pyramid) ends (Figure 3.1.7).

The pyramid of energy, therefore, gives the best picture of the relationships between **producers** (plants) and **consumers** (animals). Whereas pyramids of numbers and biomass record the organisms supported in each feeding level at any one time, the energy pyramid shows the amount of food being produced and consumed over a period of time, e.g. one year. Its shape is therefore not affected by differences in size or changes in numbers of individuals.

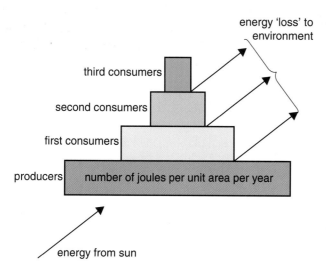

energy 'loss' to environment

third consumers

second consumers

first consumers

producers | number of joules per unit area per year

energy from sun

The energy is dispersed as it flows through a food chain

Figure 3.1.7 A pyramid of energy

Human influences on the environment

The growth of human populations is undoubtedly the main environmental influence in creating problems related to overcrowding and pollution. Factors like reproductive rates, death rates and disease all depend on the density of populations. We can determine the rate of human population increase by recording daily births and deaths. An estimated 270 000 babies are born in the world every day. There are an estimated 142 000 daily deaths. If you subtract numbers of deaths from numbers of births, you arrive at the estimated daily rate of population increase (Figure 3.1.8).

It is estimated that the human population increases by about 128 000 individuals each day. Of course, common sense tells us that this rate will probably increase. The more people

there are, the more babies there are likely to be produced. In the 21st century, the world population probably will be more than eight thousand million people. If our present rate of food production stays the same, there will not be enough food to feed everyone. Already, more than 10 000 people die from starvation or malnutrition each day.

Superimposed on this problem of a population explosion are the many problems of **pollution** that come with the use of limited space by an ever-increasing number of people.

The biggest influence we have on the environment is its destruction through pollution. A pollutant is a substance whose presence in the environment is harmful. Not only do pollutants take the form of chemicals but they also include heat and noise. Nature has always produced pollutants such as natural oil seepages, volcanic gases, and products of combustion from forest fires. These forms of pollution have always been dealt with by nature, without leaving any permanent problems. However, since the evolution of humans, the level of pollution has increased to such an extent that it now threatens our health. We have perfected our ability to pollute our environment. Indeed, we are now so good at chemistry that we can make harmful materials which cannot be decomposed. They simply accumulate progressively, providing our present generation with problems and also presenting future generations with even bigger ones.

Air, water and land are polluted daily by thousands of tonnes of pollutants on a worldwide scale. It is only relatively recently that measures have been taken to redress the balance. Unfortunately this has been too late for some species which have become extinct as a result of the destruction of their habitats. Ironically, apart from the smallpox virus, mankind has not been able to make any harmful species extinct, even though many attempts have been made using pesticides, herbicides and antibiotics.

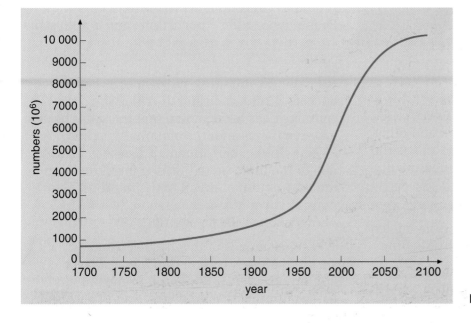

Figure 3.1.8 The population explosion

Air pollution

It has been estimated that billions of tonnes of airborne wastes reach the atmosphere each year. This is a staggering amount of pollution and has led nations to make laws banning an increase in gases produced by industrial combustion. Indeed, they aim to reduce this form of pollution to a fraction of its present amount. We all become aware of the problem when the TV weatherperson gives us an idea of air quality. In some cities, such as Los Angeles in the USA, people are advised to stay indoors and remain inactive at certain times when there are very high levels of air pollution.

Most air pollutants are products of combustion of fossil fuels. Of these, **sulphur dioxide**, is one of the most deadly. The main source of sulphur dioxide is power stations and industrial plants. Their fuels are mainly coal and oil which contain sulphur. About 65% of sulphur dioxide is produced in urban areas. The problem is that sulphur dioxide combines with rain water to form sulphuric acid.

Oxides of nitrogen are also produced from traffic exhausts and dissolve in rainwater to form nitric acid. The mixture forms **acid rain** which falls on land and water with disastrous consequences for organisms that live there. The acid causes metal ions to separate from salts

A modern power station

which occur naturally in soils. The result is that trees are killed by the toxic metal ions. Acid rain also affects water in which fish and other forms of aquatic life exist. The acid affects the gills and eggs of fish, resulting in their death.

Stonework is also corroded as acids dissolve it, causing millions of pounds worth of damage. Many famous classical sculptures from ancient civilisations have to be protected by keeping

This statue's nose has been eroded away by acid rain

them indoors, away from atmospheric conditions.

Incomplete combustion of fossil fuels produces **carbon monoxide**. Traffic exhausts are the major cause of this pollution. More carbon monoxide is released into the atmosphere from heavy traffic in cities than any other air pollutant. It combines with haemoglobin in your blood (see page 33) faster than does oxygen. So it is a deadly gas, causing suffocation.

Oxides of nitrogen combine with hydrocarbons under the action of sunlight and produce **photochemical smog**. When atmospheric pressure is high this smog settles over cities and produces conditions which irritate the lungs. People who suffer from asthma and emphysema are particularly distressed under these conditions.

All carbon-containing substances, when burned, produce **carbon dioxide**. Since the last century it has been shown that there has been a massive increase in the amount of carbon dioxide in the atmosphere. It has been largely due to the increase in burning fossil fuel and the destruction of much of the plant life that normally would use the carbon dioxide in photosynthesis (see page 186).

The consequence of this has been an accumulation of carbon dioxide as a layer around our planet. Radiation from sunlight passes through the atmosphere and heats up the sufrace of Earth but radiation reflected by Earth does not have enough energy to pass back through the layer. It has been estimated that this will result in global warming by 4°C in the next 20 years (Figure 3.1.9). It is sometimes called '**the greenhouse effect**', and if the prediction comes true, will result in massive climatic changes, causing melting of polar ice caps, with flooding of many areas near the sea.

Important air pollutants which are not products of combustion are **chlorofluorocarbons (CFCs)**. They break down the **ozone layer** which occurs around Earth. Normally, the ozone prevents cancer-inducing ultraviolet rays from the sun from reaching harmful levels. CFCs were once the only gases used in refrigerators and aerosols. They are now substituted by harmless alternatives.

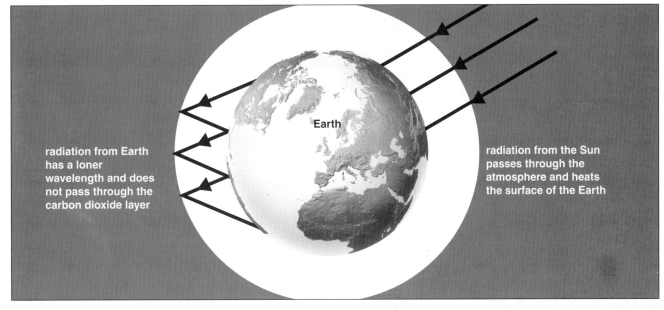

radiation from Earth has a loner wavelength and does not pass through the carbon dioxide layer

Earth

radiation from the Sun passes through the atmosphere and heats the surface of the Earth

Figure 3.1.9 Global warming

It is impossible to eliminate air pollution entirely from industrial nations but we can reduce it below danger point. **Alternative sources** of energy can help but no one pretends that wind, solar or tidal power can satisfy the ever growing energy needs of society. The greater use of **nuclear energy** would eliminate the problem of most of the polluting gases.

However, there is the problem of nuclear waste disposal which has not yet been solved (see page 168). Pollution from traffic can be reduced by the use of catalytic converters which greatly decrease sulphur dioxide and oxides of nitrogen. Lead-free petrol reduces lead pollution.

Water pollution

Many of our lakes and rivers are polluted. We have even polluted the oceans. When the forerunners of our cities were first set up, they were small settlements, often built near large rivers. One reason for this was waste disposal. Inhabitants poured untreated sewage and other refuse into the water which carried it to the sea.

It was not a serious problem at first, when settlements were small and far apart. The pollutants were decomposed by natural processes. Today, the story is completely different. Natural decomposition cannot possibly deal with the volume of waste produced by inhabitants of modern towns and cities. Strict laws are now enforced concerning the use of waterways for sewage and other waste disposal. Although **untreated sewage** is often poured into the sea, there are sewage works (see page 220) to deal with waste before

An ancient open sewer in the Lebanon

it becomes the effluent suitable to be poured into rivers.

Sewage, when allowed into waterways, will encourage bacteria to thrive. The bacteria will use all of the oxygen in the water and cause all aerobic life to die. Also, it may contaminate water with pathogens and help to spread diseases such as typhoid and cholera (see page 223). It is therefore dangerous to eat animals that come from polluted waters. Cases of typhoid, hepatitis, and other infectious diseases have often been traced to mussels or other sea life that filter water and capture the harmful bacteria.

The same problem of oxygen depletion occurs with overuse of **fertilisers** on crops. Fertiliser containing nitrate and phosphate will encourage the growth of plants when it is washed into waterways by rain. The plants reproduce to such an extent that they pile up and prevent light getting to those in the bottom layers. These die and decompose, encouraging bacteria to remove oxygen from the water.

The millions of gallons of **detergents** reaching drains daily add to water pollution problems. Many contain **phosphates** so they act as fertilisers, leading to similar problems as those posed by over-fertilising crops. Agriculture is also responsible for other forms of water pollution including those caused by **pesticides**. Again, these are washed into water and, although usually not in such concentration that is harmful to humans, becomes highly toxic as it builds up through food chains. Those animals at the top of the food chain will receive such a high concentration that they are poisoned (Figure 3.1.10).

Perhaps the most important long term form of water pollution is due to industrial wastes such as **heavy metals** like mercury, lead, zinc and cadmium. These are often by-products of manufacturing processes in paper mills, steel works, refineries and car factories. The concentration of heavy metals also increases through food chains until those feeders at the top are poisoned. The heavy metals interfere

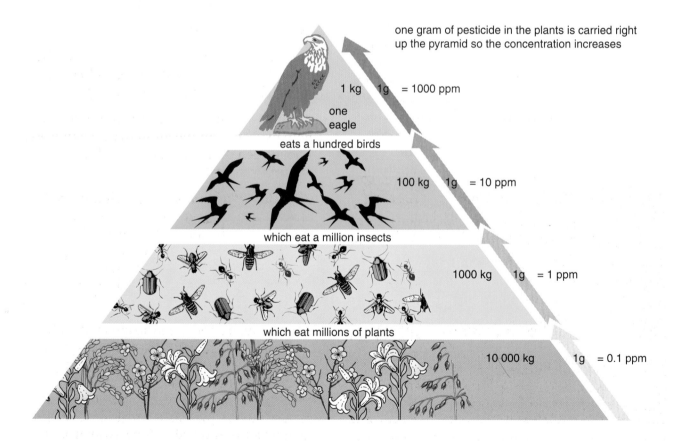

one gram of pesticide in the plants is carried right up the pyramid so the concentration increases

1 kg   1g   = 1000 ppm

one eagle

eats a hundred birds

100 kg   1g   = 10 ppm

which eat a million insects

1000 kg   1g   = 1 ppm

which eat millions of plants

10 000 kg   1g   = 0.1 ppm

Figure 3.1.10 Concentration of DDT in a food chain

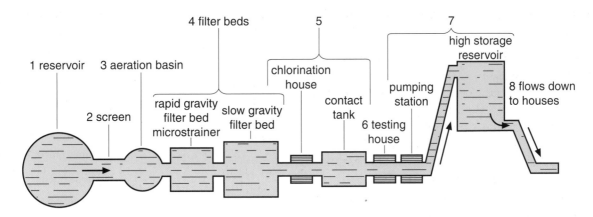

Figure 3.1.11 Water supply

with the action of many vital enzymes in animals causing death. Other non-biodegradable substances called **polychlorinated biphenyls (PCBs)** are used in the manufacture of paint and electrical equipment. There is no known method for removing them from water, and once they get into food chains, they accumulate from one feeding level to the next.

Pollution problems are also created by **oil** and other petroleum products. These come from refineries, drilling and pumping operations, shipyards and oil spills. The cumulative effects of car owners illegally dumping oil and the run off of spilled oil on roads are responsible for the large volume of oil that finds its way into water courses. Many types of wildlife, particularly sea birds, are affected directly, but the most serious problems arise when links in food chains are killed off.

Oil pollution of South Wales coast following the spill from the *Sea Empress* in February 1996

In cases of major spills, detergents used to clean up the problem are often toxic to the wildlife they are meant to save.

The problem of water pollution is not a hopeless one. The sources of pollution should be removed wherever possible and modern water treatment plants are needed to provide communities with clean water.

Water treatment plants depend on the following basic principles (Figure 3.1.11):

1 Storage in reservoirs.

2 Screening. Water passes through channels, partly blocked by metal screens. Floating material is trapped and raked away.

3 Aeration. Water is sprayed as fountains to help oxygen from the air to dissolve in water. The oxygen destroys many harmful anaerobic microbes.

4 Filtration. Water passes through filter beds made of sand and gravel. Sand and smaller stones are at the top of the filter bed and the largest are at the bottom. Water trickles slowly through the sand, and then more rapidly through the gravel and stones until it collects in pipes.

All floating matter, some suspended matter, and microbes are left behind in the spaces between sand and gravel. These primary filters are washed daily to remove impurities from the sand and gravel. Clean water and air are blown upwards through the bed and then water is allowed to flow away from the top carrying the impurities.

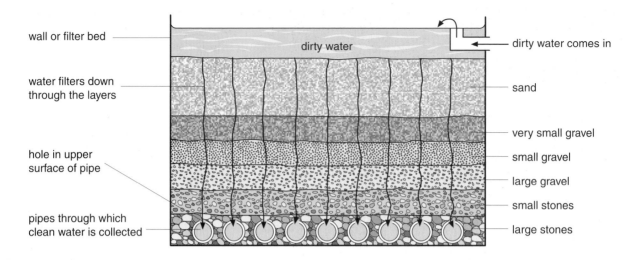

wall or filter bed

dirty water

dirty water comes in

water filters down through the layers

sand

very small gravel

hole in upper surface of pipe

small gravel

large gravel

small stones

pipes through which clean water is collected

large stones

Figure 3.1.12 Water filtration

Those filter beds which have a top layer of very fine sand allow only slow filtration. They are secondary filters. In these, impurities form a jelly-like film of microbes which kill harmful bacteria by ingesting them. Secondary filters are cleaned only once every few weeks by scraping off the top layer of dirty sand, washing it, and adding clean sand.

Modern water treatment plants filter water with micro-strainers and force water through them rather than relying on gravity filtration.

5 Chlorination. The filtered water is passed through a tank where sodium hypochlorite is added in small quantities (about 1 part to 2 million parts of water). This kills any remaining harmful microbes. Fluoride may also be added at this point to protect against dental decay.

6 The water is tested regularly thoughout the day for impurities.

7 The water is pumped to high level storage reservoirs for distribution.

Refuse and land pollution

Wherever humans go, they leave a trail of wastes. In fact, archaeology is dependent on the discovery of materials left behind by previous civilisations. There is litter in the streets, on beaches, and along roadsides. Abandoned cars, refrigerators, and other useless rubbish are left as eyesores in areas used as scrap yards. Litter is a costly and serious aspect of a waste problem that we have made for ourselves. We live in a society with more and more disposable materials being made because we depend on a constant turnover of manufactured products to create jobs and wealth.

We produce billions of tonnes of rubbish each year. It is estimated that over two kilograms of rubbish are produced per person, per day in Britain. Many people take rubbish disposal for

granted until something happens to prevent it being collected each week. Food scraps, newspapers, wood, lawn mowings, glass, cans, old appliances, tyres, furniture, and many other items make up the contents of rubbish collection trucks. There are several ways these solid wastes are disposed of. Some are dumped at sea, several miles out in the ocean. Some are burned in incinerators. Most are used as landfill.

Any method is certain to pollute the environment to some extent but landfill is probably the safest. Problems have arisen, however, when organic material is decomposed and there has been an accumulation of **methane**. Sometimes this has reached the surface and has ignited with disastrous

A landfill site

method of separating reusable wastes from the remaining wastes was needed. Today, with the reduction of natural resources, we are forced to consider recycling. However, recycling is still far from being totally accepted. Lack of public interest, recycling centres, funding for research, recyclable goods, and government incentive have slowed the process. Attitudes are beginning to change.

Environmental groups have been active in setting up collection points. The public are becoming increasingly aware of the need to take bottles, cans, old clothes and paper to be recycled. It is also up to industry to start recycling on a large scale and up to governments to propose legislation to help solve the problem.

consequences. Houses have exploded as a result. Good use is made of some of the methane produced in this way by collecting it and using it as a fuel.

One solution to the problem of rubbish is to use it over again in the process of **recycling**. It reduces the production of waste materials that are a source of pollution. It also helps save our non replacable resources such as minerals and our replaceable resources such as trees.

It is possible to recycle paper, glass, textiles and certain metals. In the past, high cost and the abundance of natural resources limited the development of recycling technology. Also an efficient

A recycling collection centre

HUMANS AND THEIR ENVIRONMENT

Summary

1 Almost all forms of life depend on photosynthesis which only occurs in green plants.

2 Through photosynthesis, carbon dioxide and water are built up into organic compounds such as glucose, the basic fuel for all cells.

3 Light energy is necessary for photosynthesis and is changed to chemical energy which is stored in glucose.

4 Food chains begin with plants and represent energy flow from one feeding level to the next.

5 Feeding relationships can be shown in the form of pyramids which demonstrate the rule that biomass decreases from the base of the pyramid to the apex because of the energy loss between feeding levels.

6 A stable population density depends on a balance between birth rate and death rate. Present trends in human populations suggest that future generations will have difficulty in providing enough food for themselves.

7 The large numbers of people living in limited space, with the consequences of industrial demands, cause pollution problems.

8 Air is polluted with many harmful gases, mainly from industrial combustion and traffic.

9 Water is polluted by chemicals including those produced as by-products of industry, fertilisers, detergents and oil.

10 Land is being used and polluted by the disposal of both biodegradable and non-biodegradable wastes.

Questions for review

1 Explain why photosynthesis is essential to almost all forms of life.

2 Define photosynthesis using a word equation and chemical formulae.

3 Starting with the food producers, name and define the various types of organisms in a food chain with five links and ending with humans.

4 Explain what you would need to know to convert a food chain into a pyramid of biomass.

5 List four types of water pollution and suggest the steps that would be necessary to solve the problem.

6 Explain how the decomposition of sewage in water kills aquatic life.

7 List four toxic gases associated with air pollution.

8 Explain how recycling can help solve waste disposal.

Applying principles and concepts

1 Give an account of the conditions that affect photosynthesis.

2 Make a diagram of a food web that exists in your local area.

3 Discuss the statement: 'Air pollution is a necessary evil'.

4 Explain how overproduction leads to mineral depletion.

5 What are the possible consequences to a wide variety of other organisms, of increased industrialisation?

HUMANS AND THEIR ENVIRONMENT

Healthy living

Learning Objectives

By the end of this chapter you should be able to understand:

- The importance of exercise and fitness to the individual
- The importance of basic standards of personal cleanliness
- The need for regular dental care, diet and weight control

- The use and abuse of drugs including antibiotics
- Personal and social implications of addiction

Good health

Health is not only the absence of illness. It is a positive and enjoyable feeling of wellbeing as a result of an effort to maintain an all-round state of mental and physical fitness. Although there is no need to be as fit as an olympic athlete, you should be fit enough to be able to live the lifestyle that you want. Our lifestyles are probably less strenuous than they were a hundred years ago, but we all have to exert ourselves from time to time, whether it is running for a bus or dancing all night. If we can do these things without collapsing, it is obviously to our advantage. Taking regular exercise has several benefits:

- Your fitness will increase.
- It reduces the chance of obesity.
- It is fun as long as you exercise in ways that suit your life and that give you pleasure.
- It will help you feel better and more self-confident.

Biologically-speaking, fitness helps the way in which your body functions. For example, sustained regular exercise improves blood circulation to your muscles, increases the stroke volume of your heart, and reduces your rate of heart beat. It also makes your breathing system more efficient. Combined with a healthy, balanced diet and a non-smoking lifestyle, exercise is certain to be beneficial to you.

Personal health care

You, personally, have more influence over your own health than any other single factor. Care of your body, clothing, eating habits, and daily activities are all under your control. Good health is largely a question of learning good habits early in life.

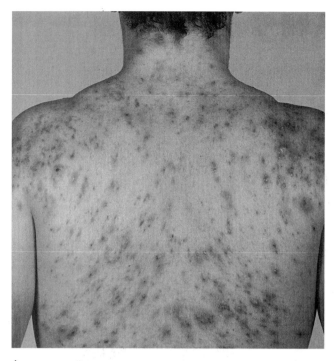

Acne

Care of skin and hair

The surface of the skin is covered with sebum from the sebaceous glands, sweat which contains salts, urea and other organic substances, bacteria, and possibly other parasites. If sebum, sweat and dirt are not regularly washed off the skin, then substances accumulate in the pores and become visible as blackheads. These may become the centre of skin infections. Impetigo and erysepelas are two infections of the skin caused by the bacterium, *Streptococcus*.

Spots, pimples and boils are usually caused by bacterial infections. Some are produced by the bacteria *Streptococcus* or *Staphylococcus*. Washing with water alone is not sufficient because it does not remove accumulated sebum. Sebum is an oil and, therefore, oil-based soaps are needed to dissolve and remove it. Shampoo for washing hair is basically a liquid soap which dissolves the sebum in the hair and removes it. Some oil is needed to lubricate the hair, but excess oil makes it greasy and encourages parasites.

A daily shower or bath is advisable because sweat and sebum collect and bacteria flourish where skin surfaces meet, such as the groin. In addition to washing, it may be helpful to use one of the many deodorants available on the market. A deodorant disguises the odour produced by the organic substances found in sweat. Particularly in hot conditions, with profuse sweating, body odour can be unpleasant and embarrassing. Care sould be taken when selecting a deodorant. Some preparations also claim to be anti-perspirants but most of these work by blocking sweat pores and this is not advisable for long periods.

A cosmetic in common use is hair dye. Some people are allergic to bleach, present in permanent hair dyes (lasting up to six months). It can cause extreme irritation. Even more serious is the evidence to suggest that some permanent dyes and toners (rinses and semi-permanent dyes) may be carcinogenic (cause cancer).

Care of the teeth

The structure of a tooth is described on page 21. Teeth grow in the warm, moist, food-filled environment of your mouth which is ideal for the growth of bacteria. Although enamel is the hardest substance in the body, it can be dissolved by acid produced when bacteria digest sugar left on the teeth. Strong enamel is a good deterrent to tooth decay. There is evidence to suggest that enamel on teeth in well-exercised jaws becomes harder. Eskimos living on their own traditional diets, do not suffer from tooth decay because they have a very low sugar diet (Figure 3.2.1).

Periodontal disease of the gums and fibrous membrane accounts for most loss of teeth in later life. Gums become soft and inflamed and shrink away from the teeth. The fibres in the fibrous membrane become damaged, break, and the tooth loosens in its socket. Plaque is the main cause of periodontal disease. It is a hard deposit which builds up on teeth and consists of a mixture of salts from salivary secretions, bacteria and food debris. Build-up of plaque on the base of a tooth causes destruction of the fibrous membrane. It also contributes to tooth decay.

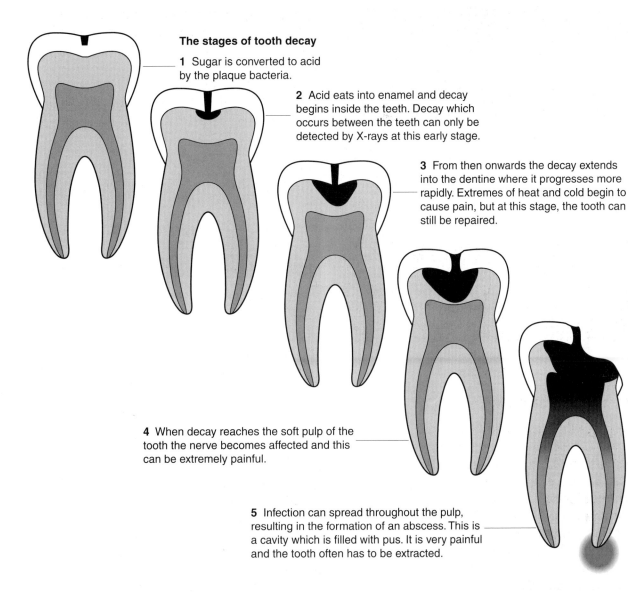

The stages of tooth decay

1 Sugar is converted to acid by the plaque bacteria.

2 Acid eats into enamel and decay begins inside the teeth. Decay which occurs between the teeth can only be detected by X-rays at this early stage.

3 From then onwards the decay extends into the dentine where it progresses more rapidly. Extremes of heat and cold begin to cause pain, but at this stage, the tooth can still be repaired.

4 When decay reaches the soft pulp of the tooth the nerve becomes affected and this can be extremely painful.

5 Infection can spread throughout the pulp, resulting in the formation of an abscess. This is a cavity which is filled with pus. It is very painful and the tooth often has to be extracted.

Figure 3.2.1 The stages of tooth decay

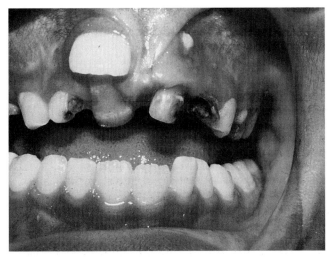

Advanced tooth decay

A micrograph of plaque on teeth

HUMANS AND THEIR ENVIRONMENT

Regular cleaning of the teeth with toothpaste and water reduces the amount of tooth decay and periodontal disease.

The toothbrush used is very important. Very soft bristles will not dislodge food particles and hard bristles could damage the gums and make them bleed. A good compromise is a brush with medium strength bristles on the outside and harder ones on the inside (Figure 3.2.2).

Toothpaste contains an abrasive substance which helps to remove the film which may produce plaque. Over-abrasive toothpaste could damage the enamel. Some toothpaste contains fluoride which could help in areas where there is no fluoride added to the water supply. Some toothpastes contain a mild antiseptic to help remove bacteria from the mouth.

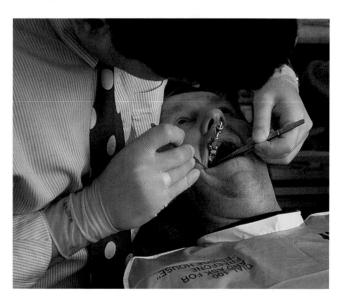

Regular visits to the dentist are important in maintaining healthy teeth

a)

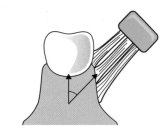

Start with the tufts of the brush pointing to the junction of the teeth and gums at an angle of about 45° to the teeth.

b)

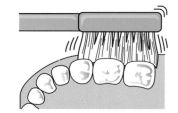

Vibrate the brush in a circular motion gently but firmly round the necks of the teeth and in between them.

c)

Brush the backs of the front teeth as illustrated.

d)

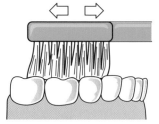

Brush the biting surfaces with a backwards and forwards scrubbing movement.

e)

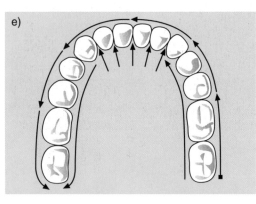

It is important to establish a routine that ensures that no area of the teeth remains unbrushed, see diagram (e). Begin at the back of the lower jaw and work methodically around the outer surfaces. Then do the same with the inner surfaces. Repeat the process for the upper jaw. Then brush the biting surfaces of both upper and lower teeth. From time to time you should check that you are successfully removing all the plaque. The only way to do this is to use a disclosing tablet.

Figure 3.2.2 How to use a toothbrush properly

A very important part of caring for the teeth is to visit a dentist regularly. In addition, many dentists employ a dental hygienist who removes plaque regularly. This particularly helps to keep gums healthy by reducing the likelihood of infection.

Tooth decay may damage the general health.

When decay reaches the root, toxins made by bacteria may set up an abscess at the base of the root. If the abscess bursts, there is a chance that the bacterial toxins will enter the bloodstream. In some circumstances, septicaemia (blood poisoning) may be the result, which was often fatal before the discovery and use of antibiotics.

Healthy eating

In order to maintain health, we all need to eat a varied and balanced diet to avoid **malnutrition** or **starvation**. The difference between these terms is that malnutrition means a lack of a balanced diet but starvation means not having enough food to provide you with the energy that you require. You can be over-weight and still suffer from malnutrition.

Everybody requires slightly different amounts of each type of nutrient (Figure 3.2.3). Guidelines have been set up by government

medical and nutritional experts concerning the amounts of different nutrients which most of us require. These are now called Dietary Reference Values and are similar to Recommended Daily Amounts (RDAs) which are part of the labelling of many foods and drinks.

The easiest way to eat a healthy diet is to ensure that we eat a variety of foods each day. For example, a balanced diet could be chosen from the foods in Table 1 (overleaf) each day.

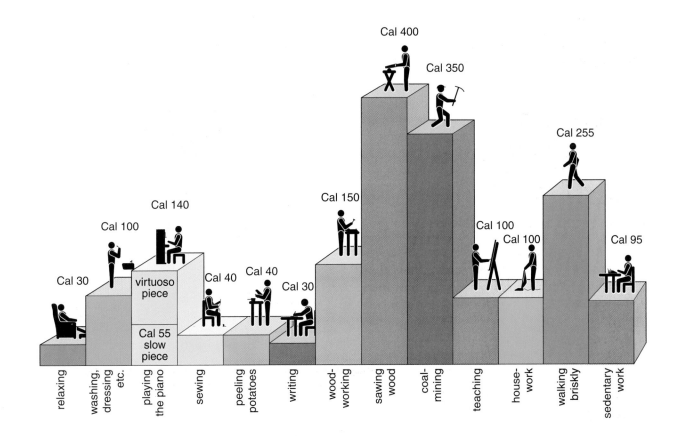

Figure 3.2.3 Energy requirements for different activities

Table I Food groups for making a balanced diet

Food group	Typical food
starchy (carbohydrate)	bread, pasta, rice, cereals, potatoes, yams, millet
dairy products (mainly fat, protein, vitamins and minerals)	milk, cheese, yoghurt
meat and meat alternatives (mainly protein)	meat and meat products, poultry, eggs, beans, lentils, nuts, soya, mycoprotein
fruit and vegetables (vitamins and minerals)	leafy, root and salad vegetables, pears, bananas, oranges, apples

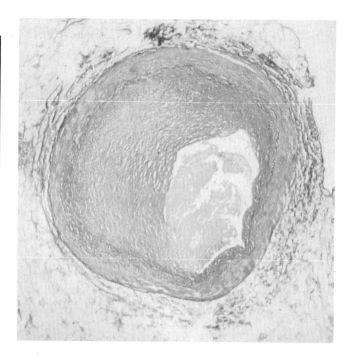

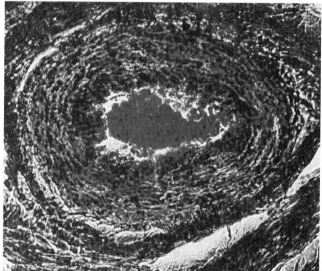

A healthy artery (top×484) and one that has become blocked by cholesterol (bottom)

Of course, there are other foods like cakes, sweets, biscuits, butter and margarine. Foods like these can be eaten as part of a well balanced diet but should not form the main part of your diet.

Starchy food will make you fat if you eat more than you need but not as easily as fatty foods. Starchy food contains only half the energy of fatty food and is much more bulky. Obesity occurs when we take more calories (energy) than we use up. Any diet which is high in fat and sugar but low in dietary fibre may be responsible for bowel cancer, heart disease and strokes. For example, too much cholesterol, a fatty substance found in all meat, can lead to a heart attack. When there is too much in the blood, it becomes deposited on the walls of arteries. This makes the central space narrower.

The coronary arteries have an enormous blood supply so the chances are that deposits will occur here. A deposit may break off and pass down the artery until it jams in a smaller artery, blocking it completely. The heart muscle is then deprived of oxygen so that its cells die and the heart stops beating.

Self-inflicted problems: tobacco, alcohol and other drugs

A **drug** is any substance which alters your physical or physiological state and the problems resulting from the use of them are not new. They have always caused social and health problems and are all harmful substances when improperly used. The facts about several drugs are given here and will form a basis for decisions concerning their use.

Tobacco – the country's leading habit

Millions of people use tobacco in some form. Most smoke cigarettes and most smokers have developed both a smoking habit and a tobacco habit. The first involves going through the motions of smoking. For example, many smokers automatically reach for a cigarette before finishing the last one. However, smokers eventually develop a physical dependence for the drug, nicotine, in tobacco and this is the tobacco habit.

Many young people who smoke feel it makes them seem more mature. Yet, if they asked people who have smoked for several years, their advice would be not to start. Unfortunately, the effects of smoking do not concern the beginner and by the time they are apparent, the beginner has become addicted.

A nationwide study took place on the effects of smoking as long ago as 1964. Some of the findings were as follows:

- Tissue damage. Lung tissue sections from thousands of smokers were examined after the smokers' deaths. Abnormal cells were found in their lungs. This was true even of those who did not die from lung cancer. Researchers also observed enlarged and ruptured air sacs and thickened arterioles.

In the trachea and bronchi, the cilia and mucus-secreting cells were destroyed. These structures normally clean and lubricate the respiratory tract and help prevent bacterial infection.

- Higher death rate. Researchers compared the number of deaths among a large sample of non-smokers with the number of deaths among a similar sample of smokers. The deaths were from many causes but certain diseases stood out. There were 1000% more deaths from lung cancer among smokers and 500% more from bronchitis and emphysema. (Emphysema is a degenerative lung disease.) The death rate was also much higher for cancer of the tongue, larynx and oesophagus, for stomach ulcers and for circulatory diseases.

- The greater the amount of smoking, the higher the death rate. In the sample, the death rate is about 40% higher for people who smoke one to ten cigarettes a day than it is for non-smokers. The death rate is 120% higher for people who smoke 40 cigarettes or more. The death rate also rises the same way with the number of years of smoking.

- Babies born to women who smoked were not as heavy as those born to non-smokers. It is

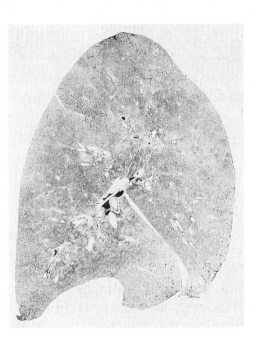

A healthy lung

A lung from someone who died from emphysema

HUMANS AND THEIR ENVIRONMENT

likely that carbon monoxide in cigarette smoke interferes with the normal oxygen-carrying ability of haemoglobin. If less oxygen is able to reach the cells, then this stunts growth.

Changing attitudes

Among more informed people, smoking has decreased since the 1960s. However, tobacco is big business and it is not in the interest of countries that rely on the tax on tobacco to ban smoking completely. Clever advertising of brands of cigarettes has maintained sales of tobacco products to young people who continue to pay to die. This is despite the compulsory health warnings on packets of cigarettes and certain restrictions on where advertising should take place.

Health education attempts by the government have failed to stop young people smoking, perhaps because it spends a small fraction on its campaigns compared to the millions of pounds spent on advertising nationwide by cigarette manufacturers.

Fortunately, more non-smoking areas have now been established in public places. This has been accepted by a majority of considerate people. Smoke irritates the eyes and throats of non-smokers and can lead to more serious passive smokers' health problems – even lung cancer – in non-smokers.

Alcohol in the body

Alcoholic drinks contain ethanol which is a poison. It is made by the action of yeast on sugars and it can be absorbed into your blood system directly through the stomach lining without being altered by digestive enzymes. Within two minutes of reaching your stomach, it starts entering your bloodstream. It is carried to your tissues and is rapidly absorbed by the cells.

In the cells, most of the ethanol is oxidised very quickly, releasing much heat which raises your blood temperature. The temperature-regulation centre of the brain is consequently stimulated and responds by causing increased circulation to the skin. Excess heat is radiated away by the increased circulation which gives a superficial rosy glow to your appearance. The rush of blood to the skin is at the expense of blood supply to the internal organs which themselves are deprived of an adequate supply of blood and heat.

Effects of ethanol on body organs

Only part of consumed ethanol is oxidised. Part is released into the lungs as vapour causing the alcoholic breath odour. Some goes to the skin and leaves in sweat. Some goes to the kidneys and leaves in urine. All body organs absorb ethanol and consequently, are affected by its presence to some extent.

The oxidation of ethanol in cells produces water and this is excreted by the skin to control the body temperature. Tissues become dehydrated and urea is concentrated in the kidneys to such an extent that it can lead to permanent kidney damage.

Excessive drinking of ethanol can also affect the stomach. It causes an increase in stomach secretions which can lead to gastritis, a painful swelling of the stomach lining.

The effects of ethanol on nerves

Ethanol is a depressant because it has an anaesthetic or numbing effect on the nervous system. However, some people mistake it for a stimulant because the numbing effect on the nerves makes some people less inhibited and less concerned about their behaviour.

The cortex of the brain shows the first effects of ethanol. Loss of judgement, willpower, and self-control are the first signs of drunkenness. When ethanol reaches the vision and speech areas of the brain, blurred vision and slurred speech result. Muscle co-ordination is affected when ethanol reaches the cerebellum (see page 74) and this becomes apparent as dizziness and the inability to walk properly. The final stage of drunkenness is unconsciousness which can become life-threatening because of the possibilty of choking on vomit or even dying because of a severely depressed heart rate.

Alcoholism – the disease

People who suffer from alcoholism depend on ethanol continually. It may start with occasional social drinking but may lead to a form of escapism to avoid stressful situations experienced because of a multitude of reasons.

About one out of ten alcoholics reaches a stage of alcohol psychosis. This is a form of mental illness requiring professional treatment and hospitalisation. Symptoms include confusion to the extent that even members of the patient's family are not recognised. Terrifying hallucinations occur involving delirium tremens, or DTs, in which uncontrollable trembling takes place.

The physiological causes may include shortage of the vitamin B complex. Indeed, vitamin deficiency diseases are common among alcoholics. This is because they often eat very little during periods of heavy drinking. As a result, the liver releases its food stores. The liver swells as the carbohydrates are replaced by fats and, over a long period, a serious liver disorder called **cirrhosis** can result. Here the fatty liver shrinks and hardens as the fats are used.

Drinking and driving – the facts

Recent controlled experiments have thoroughly tested the relationship between drinking and driving a car. A sample of experienced motorists were given a driving test similar to the one needed for a driving licence. All passed the test. The same motorists were given measured amounts of alcohol but they were not given enough for them to show any signs of drunkeness. All but one passed a breathaliser test. They were then given the same driving test again. All the drivers made many errors that could lead to accidents. Most had slower breaking reaction time. They were also inaccurate in performance but all of the drivers thought they were doing well. Their judgement was found to be impaired after only one or two drinks and this was graphically proved by psychological tests.

Therefore it is not surprising that alcohol is a factor in a large proportion of all fatal traffic accidents. In one-vehicle accidents 70% of the drivers killed had been drinking. Fifty per cent of multi-vehicle accidents are drink-related.

The facts are that alcohol

- increases reaction time
- impairs vision and distance judgement
- decreases the time span of attention
- makes it harder to associate danger signals with danger
- gives a false sense of security
- leads to aggressive driving.

Drugs and the body

The main reason why people become involved in drug abuse is because it makes them feel good. However, drugs may eventually damage the brain, liver, kidneys and lungs. Another problem is that drugs can be addictive so the user cannot stop taking them easily.

Drugs – the facts

Table 2 Types of drugs, their effects and status in the UK

Examples	Legal status	Damaging effects
alcohol	May be bought from 18 years, drunk at home from 5 years. A licence is needed to sell it.	Addictive. Slows down nervous actions (sedative). Can cause liver, brain and kidney damage and cardio-vascular problems.
barbiturates (downers, barbs)	Available on prescription only.	Drowsiness (sedative). Psychological dependence.
tranquilisers (tranx, sleeping pills)	Available on prescription only.	Drowsiness (sedative). Psychological dependence.
solvents, glue, lighter fuel	Illegal as a drug of abuse.	Dizziness, loss of co-ordination, blurred vision, nausea. Can cause permanent damage to the brain, liver and kidneys.
cannabis (pot, dope, ganja, hash, marijuana, grass)	Illegal.	Impaired judgement. Can cause lung disease through smoking. Psychological dependence. Hallucinations.
amphetamines (uppers, speed, whizz)	Available on prescription only. Illegal to sell.	Impairment of judgement and vision. Hallucinations (stimulant).
tobacco	Legal from 16 years.	Addiction. Can cause lung cancer, cardio-vascular diseases, bronchitis, emphysema.
cocaine (coke, snow)	Illegal.	Addiction (stimulant). Nervous anxiety.
cocaine freebase (crack)	Illegal.	Aggression. Feelings of fear.
caffeine (found in coffee, tea, chocolate, soft drinks)	Legal.	Stimulant.
LSD (acid)	Illegal.	Hallucinations. Can cause increased blood pressure, trembling, nausea and brain damage because of damage to nerve cells.
hallucinogenic amphetamines (ecstasy, E)	Illegal.	Hallucinations. Can cause damage to nervous system.
heroin (derivitive of opium)	Available on prescription only. Illegal to sell.	Addiction. Permanent damage to nervous system.

HUMANS AND THEIR ENVIRONMENT

Dangers of some useful drugs - antibiotics

An antibiotic is a chemical made by a microbe that can kill bacteria. No antibiotic should be used unless prescribed by a doctor for two reasons:

1 They may produce harmful side effects which may give rise to allergic reactions. In some cases the reaction is strong enough to be fatal. Certain antibiotics destroy helpful bacteria living in the intestine. Other bacteria may then grow in the intestine and cause diarrhoea or other disorders.

2 Like all organisms, bacteria show variation (see page 156). Some have varieties that allow them to resist the effects of an antibiotic and these will survive repeated use of an antibiotic. They reproduce rapidly so that a new strain may evolve in which all members resist the antibiotic.

Summary

1 Hygiene is important to healthy living.

2 The health of an individual depends on basic standards of personal cleanliness which include dental care, skin care, exercise and diet.

3 Correct oral hygiene prevents tooth decay and gum disease.

4 Attention to a well balanced diet and regular excercise prevents obesity and the health problems related to it.

5 Tobacco poses serious health problems by contributing to many diseases.

6 Ethanol is a depressant and, used excessively, leads to the health and social problems of addiction and alcoholism.

7 Misuse of drugs is harmful and may cause death to the user.

8 Drug addiction can be psychological, physical, or both.

9 Many drugs are illegal to use or sell.

10 Under some circumstances, the use of antibiotics can be dangerous.

Questions for review

1 Explain concisely how tooth decay takes place.

2 Describe the possible dangers of excess cholesterol in the diet.

3 State some of the findings concerning the death rate among smokers.

4 What are some short term disadvantages of smoking?

5 Explain the progressive effects of ethanol on the nervous system.

6 Give examples of addictive drugs which are illegal to take and sell.

Applying principles and concepts

1 Why is inhaling smoke from a cigarette injurious to your health?

2 Why is drinking alcoholic drinks on an empty stomach more dangerous than drinking after eating?

3 Why does the presence of ethanol in a person's body give the feeling of warmth?

4 Explain the possible relationship between drug addiction and juvenile delinquency.

HUMANS AND THEIR ENVIRONMENT

HUMANS AND THEIR ENVIRONMENT

Microbes and mankind

Learning Objectives

By the end of this chapter you should be able to:

■ Understand that certain bacteria are essential to life as we know it, e.g. those involved in the carbon and nitrogen cycles
■ Know that many microbes are used in biotechnology
■ Know that disease can be caused by some microbes through droplet

infection, transmission by sexual contact, contaminated food and water, and by carriers
■ Understand how the body defends itself against disease, including through immunity

The good, the bad and the ugly

It is unfortunate that when most people hear of microbes and bacteria they immediately associate them with 'germs' and disease. In fact, the vast majority of bacteria have nothing whatsoever to do with disease. Most are completely harmless; indeed, many are so

helpful to us that life as we know it would be impossible without them. It is just a relatively small number, called **pathogens**, that cause disease. This is just as well because we are constantly being colonised by millions of bacteria no matter how hygienically we live.

Essential bacteria

Certain bacteria are essential for recycling materials in our environment. One group decomposes dead things, breaking them down to chemicals which can be used by another group to start the building-up process all over again. Without these we would simply be suffocated by millions and millions of years of accumulation of dead organisms. This can be illustrated by two cycles which are essential to our lives. The first is the carbon cycle (Figure 3.3.1).

Two basic life processes are involved in the carbon cycle. These are **respiration** and **photosynthesis**. Animals and plants take in oxygen for respiration (see page 7). During respiration, glucose containing carbon is oxidised. As a result, carbon dioxide is released into the environment. If these processes continued, we would run out of oxygen and carbon dioxide would continually build up in the atmosphere. However, nature has developed

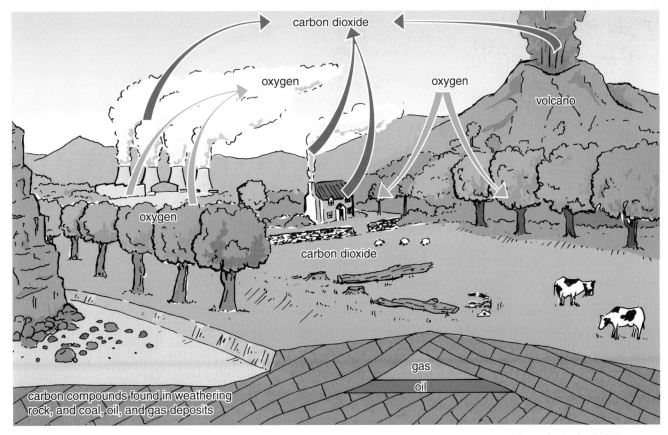

Figure 3.3.1 The carbon cycle

a neat recycling trick to keep the oxygen and carbon dioxide levels in the atmosphere relatively constant.

When plants photosynthesise they take in the carbon dioxide that would otherwise build up in the atmosphere, and they give out oxygen as a waste product. As plants photosynthesise, they make more than enough food for themselves. Plant-eating animals can take advantage of this and use the foods made in photosynthesis to make their own protoplasm. These herbivores, in turn, are eaten by carnivores and, when any organism dies, the organic materials in it are broken down by essential putrefying bacteria. The carbon leaves the dead bodies during decay as carbon dioxide. Burning fossil fuel also releases carbon dioxide into the atmosphere. The consequences of humans interfering with this fine balance are discussed on page 195.

Another cycle, involving nitrogen, also demonstrates the point that bacteria are vital for our survival.

The nitrogen cycle involves plants, animals and several kinds of bacteria. The plants' roots absorb nitrates from the soil to use them in the manufacture of proteins. So, why isn't all the nitrate in the world used up by plants? The following account should answer this question.

Herbivores obtain proteins by eating plants. Carnivores can obtain proteins by eating herbivores. Not all the protein ingested by these animals is used to build up their protoplasm. All that is ingested is digested to amino acids (see page 17), but there are more of these building blocks than the animal needs for building its own body. The excess is changed into urea which is excreted in urine.

Putrefying bacteria break this waste material down, together with the dead remains of animals and plants. Ammonia is formed but this is so reactive that it combines with chemicals in the soil to become ammonium compounds. **Nitrifying bacteria** oxidise these compounds to form nitrites and then nitrates.

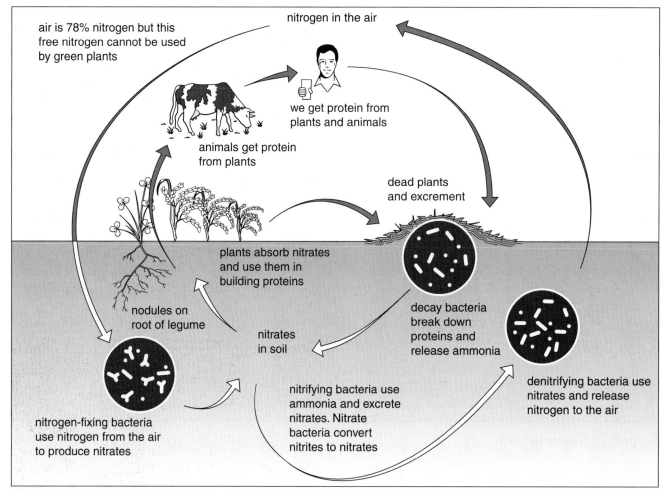

Figure 3.3.2 The nitrogen cycle

The nitrates are absorbed by plants and the process begins all over again.

About 78% of the atmosphere is nitrogen but this cannot be used by plants. However, a type of bacteria which lives in the roots of legumes can use atmospheric nitrogen and convert it into nitrates which can be used by plants. Legumes are pod-bearing plants like peas, beans, clover and lupins. The bacteria get carbohydrates from the legumes and supply them with nitrates in return. The process is called **nitrogen fixation**.

A fourth group of bacteria, called **denitrifying bacteria**, release nitrogen from nitrates in the soil. They carry out the process in anaerobic conditions (see page 8) and are most abundant in tightly-packed, water-logged soil. If the soil is well drained, loss of nitrate in this way is minimal.

Exploited microbes and biotechnology

Some microbes have been used by mankind for thousands of years to make substances that we use. We now call this **biotechnology**, the earliest form of which was probably the manufacture of bread, ethanol (alcohol), and cheese.

The microbes involved in biotechnology include bacteria, fungi, and sometimes algae. It is beyond the scope of this text to catalogue all the posible uses of these microbes, but some examples will serve to illustrate how useful some microbes are.

Useful bacteria in yoghurt production

In the production of yoghurt, a starter culture of bacteria is added to milk at 30°C. The bacteria ferment the milk sugar (lactose), producing lactic acid. This causes the protein in the milk to form a semi solid product. The bacteria used are *Lactobacillus bulgaricus* and *Streptococcus thermophilis*. The stages in yoghurt production are as shown in Figure 3.3.3.

Milk is first pasteurised (see page 225) and homogenised (thoroughly mixed). A 2.5% starter culture of bacteria is added at 45°C. Fermentation is allowed to continue for four hours at this temperature. During this time, lactose is changed to glucose, which is changed to lactic acid. After cooling, the yoghurt will keep for up to three weeks. If fruit is added, it reduces the shelf life of the yoghurt.

Useful fungi – ethanol production

Yeast is a fungus made of one cell. There are many different varieties of yeasts in nature. They grow wherever there is suitable food. For example some types live on grapes and other fruits. They form a thin 'bloom' which you can rub off with your fingers. You can see this in the photograph on page 217.

The first alcoholic drinks were made simply by leaving these wild yeasts to feed on grapes. People soon found that they could breed special kinds of yeasts which were better at this process of **fermentation**. When given food, in the form of sugar, yeast cells reproduce by budding, as shown in Figure 3.3.4.

Yeast does not need oxygen to respire. In fact it grows more slowly and makes less ethanol in the presence of oxygen. When given a suitable temperature, a food supply, and water, budding goes on rapidly, but yeast will form spores in conditions that are unfavourable for growth. In this form, yeast can survive low temperatures and drought. Yeast can be stored in dry conditions for long periods and still stay alive. We can buy 'packet' yeast which begins fermentation as soon as it is given the correct conditions.

When feeding, yeast secretes enzymes which digest the sugar in fruit juices. In the absence of oxygen, ethanol is made as a waste product. In the presence of oxygen, all the sugar is changed into carbon dioxide and water. Energy is released by this process and is used by yeast for growth.

glucose → ethanol + carbon dioxide + energy

Besides using ethanol for alcoholic drinks, we also use it in medical and in industrial technology. In countries such as Brazil, a large amount of ethanol produced from sugar is used as a fuel in specially adapted internal combustion engines for cars and buses. It has the advantage of not polluting the atmosphere like fossil fuels.

A gasohol-driven vehicle in Brazil. This particular car is being filled up with a mixture of alcohol and petrol at an ordinary filling station

HUMANS AND THEIR ENVIRONMENT

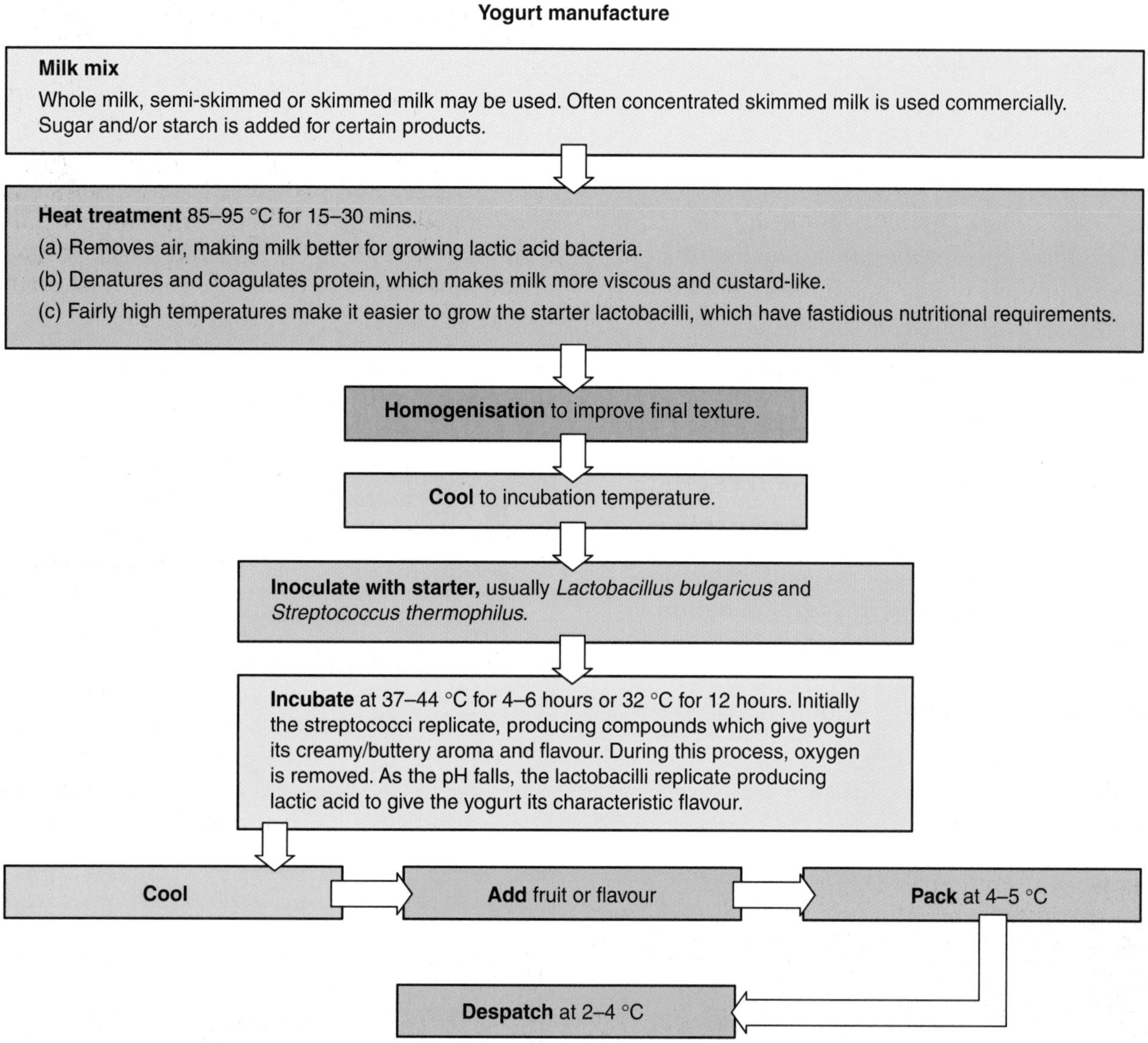

Yogurt manufacture

Milk mix
Whole milk, semi-skimmed or skimmed milk may be used. Often concentrated skimmed milk is used commercially. Sugar and/or starch is added for certain products.

Heat treatment 85–95 °C for 15–30 mins.

(a) Removes air, making milk better for growing lactic acid bacteria.

(b) Denatures and coagulates protein, which makes milk more viscous and custard-like.

(c) Fairly high temperatures make it easier to grow the starter lactobacilli, which have fastidious nutritional requirements.

Homogenisation to improve final texture.

Cool to incubation temperature.

Inoculate with starter, usually *Lactobacillus bulgaricus* and *Streptococcus thermophilus*.

Incubate at 37–44 °C for 4–6 hours or 32 °C for 12 hours. Initially the streptococci replicate, producing compounds which give yogurt its creamy/buttery aroma and flavour. During this process, oxygen is removed. As the pH falls, the lactobacilli replicate producing lactic acid to give the yogurt its characteristic flavour.

Cool

Add fruit or flavour

Pack at 4–5 °C

Despatch at 2–4 °C

Figure 3.3.3 Yoghurt production

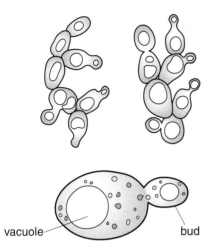

A stage in the commercial production of bread

Fungi and bread production

As you can see from the word equation on page 215, carbon dioxide is also produced when yeast acts on glucose. Use is made of this in the production of bread. The dough for bread making consists of flour, water, yeast, salt and sugar. As soon as dough is prepared, the yeast starts to feed on the sugar. Carbon dioxide is produced and makes the bread rise.

Single-cell protein (SCP)

Microbes may be used as alternatives or supplements to the normal diet of humans and domesticated animals. These microbes are rich in protein and can be used to replace traditional sources of animal and plant protein.

In 1964, research began on the possibility of increasing the production of protein by harnessing the metabolism of a fungus. *Fusarium graminearum* was the fungus and proved to be capable of providing a food which is protein- and fibre-rich. When given wheat starch as a food, water, and a suitable temperature in a fermenter, this fungus produces vast quantities of what has come to be known as mycoprotein (literally 'fungus protein'). It is sold under various brand names but the most popular is Quorn (below). It is

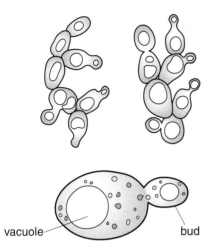

Figure 3.3.4 Yeast budding

vacuole   bud

Grapes with a yeast bloom

A Quorn product

How bread is made

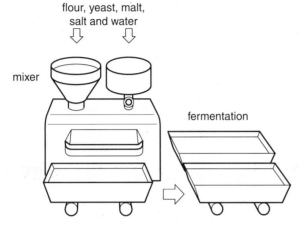

flour, yeast, malt, salt and water

mixer

fermentation

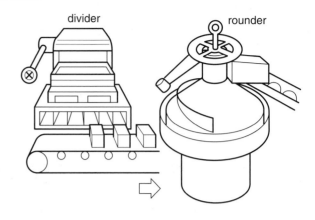

divider

rounder

1 A 'sponge' or starting mixture is made which contains only part of the total amount of flour. This is kneaded in a mixer for several minutes. It is then fermented for several hours.

2 The rest of the flour is added and the dough mixed again. A divider cuts the dough into loaf-sized pieces. These are shaped by the rounder.

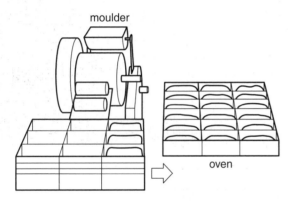

moulder

oven

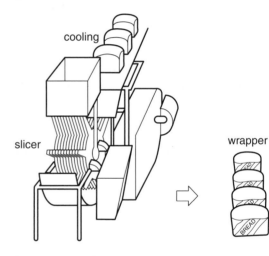

cooling

slicer

wrapper

3 A moulder shapes the pieces of dough into cylinders that are put into baking pans. The dough ferments again. It is then put in the oven.

4 The loaves are baked for 20 minutes. They are cooled, sliced and wrapped.

Figure 3.3.5 Bread production

HUMANS AND THEIR ENVIRONMENT

collected from the fermenter and, after pasteurisation, is processed into food products, especially pies and curries. Compared to lean steak it has the composition shown in Table 1.

Although lean steak is higher in protein, it has the disadvantages of having a lot of fat and no fibre. Problems related to cholesterol intake and lack of fibre are reduced by eating mycoprotein.

Table 1 The dietary composition of mycoprotein and lean steak

	percentage weight without water	
	mycoprotein	**lean steak**
protein	44.3	68.2
fat	13.8	30.2
fibre	37.6	0

Antibiotics

As long ago as 1939, mass production of the antibiotic, penicillin, made by the fungus *Penicillium* was attempted. Although it was Alexander Fleming who discovered the antibiotic in 1928, the pioneers in mass production were Howard Florey and Ernst Chain several years later.

In the production of penicillin, the starter culture of *Penicillium* is added to a liquid culture medium in a fermenter. During the first 24 hours, the cells multiply rapidly. When the sugar in the medium decreases, the fungus produces penicillin. After seven days, the

Mass production of penicillin in the 1940s

Alexander Fleming (1881–1955)

Howard Florey (left) and Ernst Chain (right)

concentration of penicillin in the medium reaches its maximum. The medium is filtered and the penicillin is extracted from the filtrate.

The problem that had to be solved in mass production was that the fungus uses up all its food and air as it grows. It produces waste

products that build up and harm it. Biotechnologists designed production units which kept adding new food and air while removing waste products. Giant fermenters are used for this process today with capacities of over 1000 gallons.

Microbes and waste disposal

About 400 litres of sewage are produced per person, per day in the UK. Only about half a litre of this is the solid which could cause a health risk. The problem stems from the variety of bacteria which live on the solids. Many are harmless to humans but can still cause

pollution (see page 195). Because of this, untreated sewage can no longer be legally dumped into water sources in Britain. It must be treated at a sewage works so that the organic matter in the material leaving the works (**effluent**) is minimal.

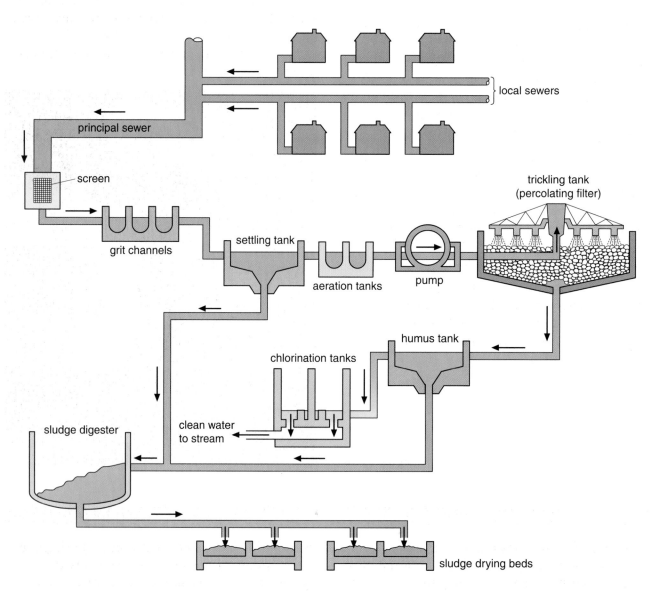

Figure 3.3.6 Sewage works

Microbes are involved in three processes in sewage treatment:

1 biological filtration

2 activated sludge

3 anaerobic digestion.

The helpful microbes responsible during the treatment are:

- Bacteria which digest down cellulose (fibre in sewage).
- Fungi which break down industrial wastes in acid conditions and destroy nematode worms in the sewage.
- Algae which provide oxygen. This helps to kill harmful bacteria which live without oxygen.

Biological filters usually consist of a tank filled with inactive materials such as small stones. Liquid effluent trickles over the stones, on which there is a film of aerobic microbes. These microbes feed on the organic matter in the effluent and bacteria convert ammonium compounds in the effluent to nitrates.

In **activated sludge tanks**, oxygen is added to sewage by pumping air through. Microbes digest organic matter from the sewage and the liquid is then led to sedimentation tanks. Here, any remaining organic matter settles to the bottom and the water leaving these tanks is often clean enough to be discharged into rivers.

In **anareobic digesters**, bacteria digest the sludge which remains. During this process, methane gas is produced which is often used as a fuel to release energy for the activities of the sewage works itself.

Biogas – fuel from microbes

When bacteria are used in the anaerobic digestion of sewage, methane is produced. The name, **biogas**, is given to this because it is produced from rotting plant and animal material.

decaying organisms → methane + energy

In China and India, rotting vegetation and animal dung are fermented to produce biogas for use as a fuel for families. Indeed, many families in rural districts in the Far East have biogas generators attached to their houses. The gas produced is used for cooking and lighting on a relatively small scale. In Britain, refuse dumps produce large amounts of methane, but most of it is wasted. However, in certain cities, methane is collected and piped away from refuse dumps to be used as a fuel for hospitals and other larger buildings.

A biogas generator

Harmful microbes – the bad and the ugly

Those helpful and essential organisms described previously are probably not the most talked about microbes. The ones that get their names in newspapers and on TV and radio are usually the disease-causing types – **pathogens**. They can be killers and, because many can spread from one person to another so easily, many are the cause of some important diseases.

How do pathogens cause disease?

Like all living things, pathogens produce waste products. Unfortunately for the infected person, these act like poisons and can kill cells. If the microbes are able to establish themselves and thrive within us, they overcome our natural defence mechanisms and so produce even more chemicals to poison our organ systems. We then show symptoms of the disease. However, there are ways that we try to defend the body against invasion by microbes:

- Skin acts as a barrier.
- Tear glands produce an anti-bacterial chemical.
- Blood clotting provides a temporary barrier before a wound heals.
- White blood cells ingest microbes.
- The stomach produces hydrochloric acid which sterilises food.
- Antibodies and antitoxins are produced.

These are some of the methods that have evolved in our bodies to help our survival despite continued attack from microbes. Once established in or on our bodies, many can spread to other people

- through the air
- by direct contact
- in body fluids such as blood and semen
- through food and water
- by carriers such as insects and other animals.

Some of the culprits

Airborne diseases include influenza and tuberculosis. **Influenza** (flu) is caused by a virus which normally enters the body through the respiratory passages. It attacks the membranes which line these passages and, in doing so, will damage cells so that they can be invaded by bacteria causing secondary infections. Conditions of overcrowding and poor ventilation will help to spread the disease because millions of the virus can spread in airborne droplets through sneezing and coughs. There are so many different strains of this virus evolving continuously that vaccination is of limited and temporary value (see page 227).

Tuberculosis is caused by a bacterium. The bacteria enter the body through the respiratory passages and may reach many parts of the body via the bloodstream. Infection of the lungs is most common and the disease is more likely to develop if a person is living in overcrowded conditions and has a poor diet. Vaccination and mass X-ray programmes to detect early stages have helped to prevent the spread of the disease. Also, antibiotics have been successful in curing infected people. However, recently, antibiotic-resistant strains of the tuberculosis bacterium have evolved and are proving very difficult to eliminate. An increase in social deprivation, even in developed countries has also increased the incidence of tuberculosis in recent years, with homelessness and poor diet aggravating the situation.

Athlete's foot is spread by **direct contact**. It is caused by a fungus which lives off dead superficial skin cells, often beween toes. It is spread as spores which are produced by the fungus when it reproduces. The spores can reach the skin of people who share communal bathing facilities or showers. Shared towels can also spread the disease. Careful attention to foot hygiene can prevent its spread and it can be cured with creams and powders containing a mixture of fungicides and antibiotics.

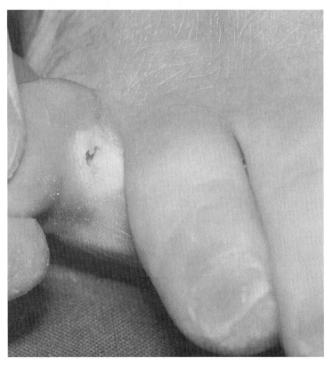

Athlete's foot

AIDS (Aquired Immune Deficiency Syndrome) is caused by **HIV (Human Immunodeficiency Virus)** and is spread through contact with body fluids such as blood and semen. It is a condition which interferes with the body's normal immunity to disease. The spread of the disease has become one of the most serious social problems in recent history. The World Health Organisation (WHO) believes that 10 million people throughout the world are infected with HIV and that over 100 000 people show symptoms of AIDS.

HIV is transmitted by all forms of unprotected sex, contact with blood (for example, the use of infected needles), and during pregnancy from mother to child. AIDS is already a worldwide tragedy. At present there is no cure, so methods of controlling its spread depend on prevention through responsible attitudes towards sexual behaviour and to the use of injected drugs.

Infected people have their natural immune systems destroyed and so become vulnerable to many different infections, such as pneumonia, hepatitis, tuberculosis, many fungal infections and herpes. They are also likely to develop forms of cancer and brain damage.

Gonorrhoea and **syphilis** are other **sexually transmitted diseases (STDs)** but can be cured with antibiotics if treated in their early stages. Symptoms of gonorrhoea in the early stages are localised as a burning sensation when urinating followed by pus in the urine. If left untreated, however, arthritis, heart disease and blindness can result. Prevention is only possible by avoiding sexual intercourse with infected people.

The symptoms of syphilis are more widespread. A sore or blister may occur on the penis or in the vagina followed by a rash, sore throat and fever after three to six weeks. Ulcers appear in the mouth, lips, and on the opening to the vagina. If the disease remains untreated, the bacteria may affect the heart, blood vessels and brain. Prevention is only possible if intercourse with an infected person is avoided.

Diseases caused by bacteria which are transmitted via food and water include cholera, salmonella and botulism.

The water supply in a village in Malawi, Africa

Cholera is spread through drinking water which has become contaminated with infected faeces. The bacteria infect the large intestine and cause severe diarrhoea. Fluid loss can be so great that the person may die of dehydration. Treatment involves replacement of fluid with saline solution and also requires antibiotics. Spread of the disease can be prevented by not contaminating water supplies with faeces and by having proper chlorination of drinking water supplies. Immunity by vaccination against the disease usually lasts for only six months.

Salmonella bacteria cause certain types of food poisoning. Food may be contaminated with bacteria from an infected person's faeces. The bacteria infect the intestine and their toxins may produce a fever followed by diarrhoea and vomiting. Careful attention to hygiene in the preparation and storage of food helps to prevent the spread of food poisoning. Also, complete thawing of frozen food before cooking is essential with no refreezing of food after it has thawed.

Canned food that has not been properly sterilised is an ideal home for a bacterium that produces one of the most deadly poisons known and which causes **botulism**. Less than a millionth of a gram will kill you. Normally the bacterium, called *Clostrium botulinum*, lives in the soil but reproduces only in anaerobic conditions. If canned food is contaminated it

This dog has rabies and must be handled with great care

A housefly feeding on bread and honey

will produce a foul-smelling gas which causes cans to swell (sometimes called 'blown' tins). This is why it can be suicidal to taste suspect food from cans. Symptoms appear about three days after eating poisoned food and include vomiting, constipation and paralysis.

Finally, there are microbes which are transmitted in or on the bodies of **carriers**. **Rabies** is still one of the most feared diseases. It is caused by a virus and transmitted by the bite of an infected animal.

All carnivores are capable of carrying the virus. It affects the nervous system and is almost invariably fatal. At present, no animals in Britain are known to be infected with the disease. There are very strict quarantine laws to prevent infected animals from entering this country.

Probably the most common insect carrier of disease in Britain is the housefly. It feeds on contaminated faeces or meat and then regurgitates it. If it comes in contact with food for human consumption, it is not surprising that many forms of harmful microbes can be spread. It spreads **dysentery** and **diarrhoea** very commonly in the summer. Strict hygiene when storing and handling food is essential to prevent the spread of disease by houseflies. With all of our sophisticated technology, we have failed to eliminate the housefly because it has evolved strains that are resistant to insecticides. The housefly's rate of evolution is faster than our rate of invention!

Control through prevention

Bacteria are among our chief competitors for food. We cannot even estimate the amount of food spoilage by bacteria but most food would remain edible for years if bacteria were not present on it. For this reason we have developed efficient methods of preserving food.

They all depend on either eliminating bacteria themselves or removing one or both of the conditions necesary for their growth – water and a suitable temperature. Killing all bacteria present, then sealing the food in a container is the basis of canning. Another method involves keeping food in conditions under which bacteria reproduce more slowly or cannot reproduce at all. This includes cooling or freezing, salting and dehydration. Chemical preservatives may also be used. However, their use has declined in recent years.

Radiation may be an important aid in preserving foods. The foods are packaged and sealed, then exposed to gamma radiation (see page 169) to destroy all bacteria.

A special form of treatment is given to milk.

Because it is an ideal food, milk can be a breeding ground for many types of bacteria which cause disease. For example, tuberculosis, diphtheria, scarlet fever, food poisoning, and dysentery can all be spread via contaminated milk unless it is treated by **pasteurisation**. This is a legal requirement before milk can be sold.

There are two methods:

1 The HTST method, which means high temperature, short time. Here milk is heated to 72°C, held at that temperature for 15 seconds, then cooled to about 10°C.

2 The milk is heated to a temperature between 63°C and 66°C, held at that temperature for 30 minutes, then cooled rapidly to about 10°C.

Pasteurisation does not eliminate all bacteria from milk but does destroy pathogenic types. Sterilised milk is also available in which the milk is heated to very high temperatures (Ultra High Temperatures, UHT). Here almost all bacteria are killed but, as a consequence, the flavour is altered. The shelf life of such milk is much longer than that of pasteurised milk.

The cellular level of defence

If microbes are able to pass through our first line of defence such as skin, and mucous membranes, they meet a second line of defence in the tissues of the body. Certain white blood cells (see page 33) can engulf and digest the invaders.

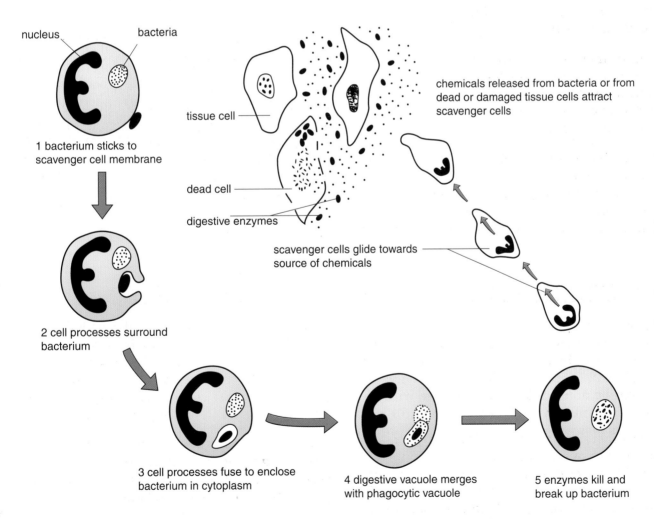

nucleus   bacteria

1 bacterium sticks to scavenger cell membrane

2 cell processes surround bacterium

3 cell processes fuse to enclose bacterium in cytoplasm

4 digestive vacuole merges with phagocytic vacuole

5 enzymes kill and break up bacterium

tissue cell

dead cell

digestive enzymes

chemicals released from bacteria or from dead or damaged tissue cells attract scavenger cells

scavenger cells glide towards source of chemicals

Figure 3.3.7 Phagocytosis

HUMANS AND THEIR ENVIRONMENT

Antibodies produced by the host's body attach to the antigens on the invading bacterium. The bacterium then is easily engulfed by a white blood cell.

white blood cell

antigens   bacterium

antibodies

a)   b)   c)   d)

Figure 3.3.8 Antigen/antibody reaction

As the battle goes on in a wound, a fluid of digested bacteria, broken down cells and blood plasma builds up as pus. Often the tissues become inflamed as increased blood flow to the area brings more white cells. Lymph (see page 48) also seeps into the infected region. It carries bacteria and white cells to lymph nodes, where they are filtered out.

A third line of defence involves production of special proteins called **antibodies** which are specific to a particular disease-causing organism. Invading microbes and their toxins have their own proteins that are alien to the host. These foreign proteins are called **antigens** and can be recognised by body cells. They cause the host's lymphocytes (see page 32) to produce specific antibodies. The antibodies combine with the antigens because they fit precisely together due to their molecular shapes (Figure 3.3.8).

Production of specific antibodies in the lymph nodes begins within a few hours of an antigen appearing in the body. Within a few days the antibodies enter the blood. They increase in number over three or four weeks, then the rate of production slows down. If the host is again exposed to the antigen, production of antibody rises very rapidly and the host may not develop any symptoms.

Organ transplants

Several kinds of organs can now be transplanted from one person to another. Probably the most common is the kidney transplant. Although the surgery may be successful, another problem often occurs. This is the problem of **rejection**, which is linked to the antigen-antibody reaction described above.

Unless the donor is an identical twin, with identical proteins, the donor's organs will have proteins that will cause the recipient to produce antibodies to destroy them. As far as the recipient's defence mechanism is concerned, it is being invaded by a foreign body.

The problem has been partially solved by the development of drugs which surpress the

A young patient one year after a successful heart transplant operation

production of antibodies by the recipient. The danger is that the recipient is now defenceless against disease-causing organisms and so has to be kept in perfectly sterile conditions until the transplanted organ has been completely accepted.

Built-in resistance to disease

Immunity is the ability of the body to resist an infectious disease. It may be present at birth, or it may be acquired during your lifetime.

We have **natural immunity** to most diseases that affect other animals and plants. Conditions in the human body are simply not suitable for most disease-causing microbes which affect other species. Fortunately for us, most pathogens are highly specific. Exceptions include tuberculosis, anthrax and rabies, which may be contracted by humans from other mammals.

Acquired immunity may be active or passive. **Active immunity** may be acquired naturally as a person recovers from certain infections. During these infections, the body builds up antibodies and these continue to be formed after recovery. This gives permanent immunity. Diphtheria, scarlet fever, measles and mumps are some diseases that usually result in this type of immunity.

Active immunity may also be acquired artificially through the use of vaccines. A vaccine contains antigens derived from pathogens which on injection, will give protection against invasion by the live pathogen.

There are several ways of producing a vaccine:

1 Using the killed pathogen e.g. whooping cough.

2 Using a weakened (**attenuated**) strain of a pathogen. The pathogen is weakened after many generations of selective sub-culturing. This can be done in the laboratory by transferring successive generations of a pathogen through a series of living cells, e.g. tuberculosis, rubella (German measles).

3 Changing the toxin so that it is no longer toxic but still has the molecular shape of the toxin's antigen to fit into an antibody, e.g. diphtheria and tetanus.

4 Isolating antigens from the pathogen and injecting them e.g. influenza.

5 Genetically engineering bacteria to mass produce antigens, e.g. hepatitis B. Vaccines cause the body to produce antibodies just as the actual infection would but the person does not suffer the symptoms of the disease.

Passive immunity involves introducing antibodies into a person. It can be acquired naturally or artificially but, in either case, it is temporary. An embryo can obtain antibodies through the placenta of its mother by diffusion from the blood. A new born baby can also receive antibodies in its mother's milk. This sort of passive immunity lasts between six months and one year.

Artificial passive immunity is acquired when a person recieves serum containing antibodies, produced by other animals. Immunity of this type is effective from a few weeks to several months.

Summary

1 Most microbes are not harmful to mankind. Some are essential, for example, those involved in natural cycles such as the carbon and nitrogen cycles.

2 Many microbes are used by us to make useful products. These inculude types involved in the production of various foods, ethanol, antibiotics and biogas.

3 Microbes are also used in sewage treatment.

4 Disease-causing microbes are called pathogens and can be transmitted through the air, by direct contact, through food and water, and by vectors (carriers).

5 Control of disease depends on hygiene, including handling and storage of food, responsible sexual attitudes and vaccination.

6 The body has natural defence mechanisms including natural immunity.

7 Natural defence mechanisms may lead to rejection of transplanted organs.

8 Artificial immunity may be acquired through immunisation.

Questions for review

1 Give an account of a food manufacturing process that depends on one or more species of microbe.

2 What do you understand by the term 'biotechnology'?

3 Describe the various ways in which infectious diseases are spread.

4 How does food poisoning occur? Give one example.

5 List the main structural defences of the body.

6 Describe how white blood cells defend the body against disease.

7 What is an antibody?

8 How do vaccines protect the body from a particular disease?

9 Distinguish between natural and artificial immunity.

10 How are antibiotics different from other drugs used in medicine?

Applying principles and concepts

1 Justify the statement that 'humans could not survive without microbes.'

2 Discuss immunity in terms of future possible uses in the treatment of HIV.

3 What is the major problem with organ transplants? What can be done to solve this problem?

4 Explain the idea that natural selection produces bacteria resistant to antibiotics.

Sample examination questions

1 In a natural environment the vegetation supports the animals and the natural cycles replenish the soil and atmosphere, thus ensuring continued plant growth. Generally, the populations in an area remain fairly constant in size from one generation to another. The birth rate, death rate, emigration (individuals leaving), and immigration rate (individuals arriving) all influence the size of the population.

Humans, however, through careless treatment of the environment, have often upset the natural balance to make way for agriculture. Massive deforestation has resulted in environmental change.

a) describe how a natural cycle restores the correct amount of carbon dioxide in the atmosphere.

b) State how the massive deforestation could be a factor influencing the greenhouse effect.

c) Besides deforestation, state two other contributors to the greenhouse effect.

d) From the passage above, use the symbols B to represent the birth rate, D the death rate, E the emigration rate and I the immigration rate, and copy and complete the equation below to indicate a population that has remained constant.

$$+ \quad = \quad +$$

WJEC, 1995

2 Consider the data in the table below concerning human populations for 1970 and 1988. Doubling time is the time taken for the population to double in size. Birth rate is the number of births per thousand population.

a) Which generation's birth rate is not following the general trend?

b) Which regions had the highest populations below puberty in 1988?

c) Which region's population was most affected by birth control in 1988?

d) i) Which region would have the largest population in the year 2000?
   ii) Suggest two factors which this region would need to develop to cope with a large population increase.

WJEC, 1995

3 Concern is rising over the effects of paints containing tin which are used on sea-going boats to prevent attachment of animals and plants on their hulls. There is evidence that high concentrations can be harmful to humans. Scientists have warned that salmon reared in fish farms can have the poisonous tin compound in their flesh. The tin-based paint is used on the enclosures where the fish are farmed. Cooking does not destroy the compound or remove it from the fish.

Discuss the statement: 'Paint containing tin used in water can be compared to pesticides used on land.'

Region	Year	Birth rate	Death rate	Doubling time of population (years)	% below 15 years of age
Africa	1970	47	20	27	44
	1988	44	15	24	42
Europe	1970	18	10	88	25
	1988	13	10	266	21
United States	1970	18	10	70	30
	1988	16	9	99	22
Russia	1970	18	8	70	28
	1988	20	10	68	26
Mexico	1970	44	10	21	46
	1988	30	6	29	42

In your discussion, write about
  i)   the advantages and disadvantages of using the paint
  ii) a possible solution to the problem in fish farms
  iii) the effects on food chains in the sea
  iv) the comparison with pesticides.

WJEC, 1995

4 The diagram shows a fermenter used to produce penicillin in large quantities from the fungus *Penicillium*.

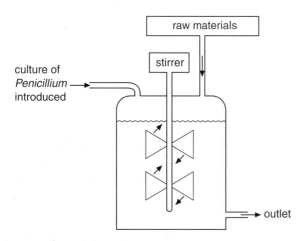

a) i) State two raw materials which *Penicillium* needs and explain why the fungus needs them.
     ii) Suggest why the mixture is stirred continuously.

b) The graph shows the production of penicillin by the fungus.

  i) After how many hours does *Penicillium* start to produce penicillin?

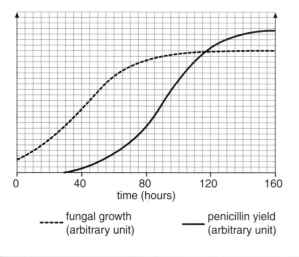

----- fungal growth (arbitrary unit)   ——— penicillin yield (arbitrary unit)

ii) Suggest a reason for the decrease in the rate of growth of *Penicillium* after 80 hours.

c) State how the fungus is separated from the penicillin at the end of production.

WJEC, 1995

5 a) Families living in rural areas of China have used biogas generators for many years. The generators are tanks made from soil and a weak cement mixture which sometimes allows the tanks to crack. The tanks are buried to reduce heat loss. Every few weeks, manure and household wastes are added to the tanks which need to be quite large. Methane is collected through pipes from the tanks. The fermented material is rich in nitrogen and is used as a fertiliser. About twice a year the tanks are cleaned but a little of the fermented material is always left at the bottom of the tank when new waste is added.

  i) State and explain one advantage and one disadvantage of using this type of biogas generator.
  ii) Explain why some fermented material should be left in the tank when it is cleaned out.

b) Chinese technologists have developed a new type of biogas generator. Study carefully the diagram opposite which shows its design.

This design makes the temperature inside the tank 10°C higher than in the old type of tank. As a result, the new generator produces methane at a faster rate than in the older design. The tank is also smaller than the older design. Technologists think that the new type of generator will be more popular and because of its small size will be possible to be used in towns.

  i)   Explain how the tank could increase the temperature inside it.
  ii) State why a higher temperature inside the new tank could improve its efficiency.
  iii) Above a certain temperature methane is not produced. Give an explanation for this.

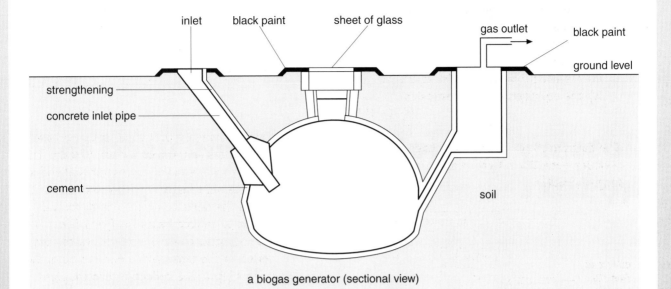

a biogas generator (sectional view)

Labels on the diagram:
- inlet
- black paint
- sheet of glass
- gas outlet
- black paint
- ground level
- strengthening
- concrete inlet pipe
- cement
- soil

6 Whooping cough is an extremely infectious disease which can lead to brain damage and even death. The whooping cough vaccine is made from dead, whole bacteria and has been used in the UK since 1949. In the mid 1970s many parents refused to have their children vaccinated because of a few highly publicised cases of brain damage and death caused by a batch of faulty vaccine. As a result, two epidemics of whooping cough followed. Today, 90% of parents choose to have their children vaccinated and the number of cases has fallen again.

In the early 1980s Japanese scientists developed a vaccine based on purified fragments of the bacterium which causes whooping cough. This 'acellular' vaccine does not cause the fever and convulsions which are sometimes caused when the 'whole' bacterium vaccine is used.

a) State four factors which influence parents in decisions as to whether to have their children vaccinated against whooping cough.

b) Besides the whooping cough vaccine, state four other ways of producing vaccines and for each, state a disease which the vaccine prevents.

c) To which type of vaccine is the Japanese vaccine, mentioned above, most similar?

WJEC, 1995

Glossary of biological terms used in the text

A

Abdomen. The area between the thorax and the pelvis.

Actin. Slender filaments of protein arranged in bundles in the composition of muscle fibrils.

Active transport. The passage of a substance through a cell membrane requiring the use of energy.

Adaptation. The process in which a species becomes better suited to survive in an environment.

Addiction. The body's need for a drug that results from the use of the drug.

Adrenal glands. Two ductless glands located above each kidney.

Aerobic. Requiring free atmospheric oxygen for normal activity.

Aerosols. Tiny suspended droplets in air.

Air sacs. Thin-walled divisions of the lungs.

Alimentary canal. Those organs that compose the food tube.

Allele. One of a pair of genes responsible for contrasting characters.

Alveoli. Microscopic sacs in the lungs in which exchange of gases takes place.

Amino acids. Substances from which organisms build protein; the end products of protein digestion.

Amnion. The innermost fetal membrane, forming a sac surrounding the fetus.

Amniotic fluid. Secreted by the amnion and filling the cavity in which the embryo lies.

Anaerobic. Deriving energy from chemical changes other than involving oxygen.

Antibiotic. A bacteria-killing substance produced by a microbe.

Antibody. A substance in the blood that helps lead to immunity.

Antigen. A substance, usually a protein, which when introduced into the body stimulates the production of antibodies.

Antitoxin. A substance in the blood that counteracts a specific toxin.

Anus. The opening at the posterior end of the alimentary canal.

Aorta. The largest artery in the body which leads from the heart.

Aquatic. Living in water.

Aqueous humor. The watery fluid filling the cavity between the cornea and the lens of the eye.

Arteriole. A tiny artery that eventually branches to become capillaries.

Artery. A large muscular blood vessel that carries blood away from the heart.

ATP (adenosine triphosphate). A high-energy compound found in cells that functions in energy storage and transfer.

Atrioventricular valves. The heart valves located between the atria and ventricles.

Atrium. A thin-walled upper chamber of the heart that receives blood from veins.

Auditory nerve. The nerve leading from the inner ear to the brain.

Autonomic nervous system. A division of the nervous system that regluates involuntary actions in internal organs.

Autosome. Any paired chromosome other than the sex chromosomes.

Axillary buds. Buds formed in the angle between a leaf stalk and a stem.

Axon. A nerve process that carries an impulse away from the nerve body.

B

Bacteria. A group of microscopic organisms without nuclear membranes.

Bile. A brownish green emulsifying fluid secreted by the liver and stored in the gall bladder.

Biodegradable. Material that can be decomposed by natural processes.

Biomass. The mass of living material per unit area or volume.

Blastula. An early stage in the development of an embryo, in which cells have divided to produce a hollow sphere.

Bowman's capsule. The cup-shaped structure forming one end of the tubule and surrounding a knot of blood capillaries in the tubule of a kidney.

Brain stem. An enlargement at the base of the brain where it connects to the spinal cord.

Breathing. The mechanism of getting air in and out of the lungs.

Bronchiole. One of numerous subdivisions of the bronchi within a lung.

Bronchus. A divison of the lower end of the wind pipe leading to a lung.

C

Calorie. A unit used to measure the energy in food. It is now outdated and should be substituted by the joule as a unit. A calorie is the amount of heat needed to raise the temperature of 1 gram of water (1 millilitre) one degree Celcius.

Canine. Teeth for tearing.

Capillary. The smallest blood vessels in the body through which exchanges occur between blood and tissue fluid.

Cardiac muscle. Muscle composing the heart wall.

Carnivore. A meat-eater.

Cartilage. A strong, pliable, smooth tissue which supports structures and lines bones at joints.

Catalyst. A substance that accelerates a chemical reaction without being altered chemically.

Cell. A unit of structure and function of an organism.

Cementum. The covering of the root of a tooth.

Central nervous system. The brain and spinal cord, and nerves arising from them.

Cerebellum. The brain region between the cerebrum and medulla, concerned with balance and muscular co-ordination.

Cerebrum. The largest region of the brain, considered to be the seat of emotions, intelligence, and voluntary activities.

Cervix. The neck of the uterus.

Chlorophyll. Green pigment essential to food manufacture in plants.

Chloroplast. A structure in a cell that contains chlorophyll.

Choroid layer. The layer of the eye beneath the sclera which prevents internal reflection and contains most blood vessels.

Chromatid. During cell division, each part of a double-stranded chromosome.

Chromosome. A rod-shaped, gene-bearing body in the cell, composed of DNA joined to protein molecules.

Cilia. Tiny, hair-like projections of cytoplasm.

Ciliary muscles. Those that control the shape of the lens in the eye.

Cleavage. The rapid series of divisions that a fertilised egg undergoes.

Cochlea. The hearing apparatus of the inner ear.

Coenzyme. A molecule that works with an enzyme in catalysing a reaction.

Cone. Cells in the retina responsible for colour-vision.

Colloid. A gelatinous substance, such as protoplasm or egg albumen, in which solids are dispersed throughout a liquid.

Colon. Part of the large intestine.

Connective tissue. A type of tissue that lies between groups of nerve, gland, and muscle cells.

Cornea. A transparent bulge of the sclera of the eye in front of the iris, through which light passes.

Corpus luteum. Refers to the follicle in an ovary after an ovum is discharged.

Cowper's gland. Located near the upper end of the male urethra. It secretes a fluid which is added to the sperms.

Cytoplasm. The protoplasmic materials in a cell lying outside the nucleus and inside the cell membrane.

D

Daughter cells. Newly-formed cells resulting from the division of a previously existing cell, called a mother cell. The two daughter cells receive identical nuclear materials.

Decomposers. Organisms that break down the tissues and excretory material of other organisms into simpler substances through the process of decay.

Deficiency disease. A condition resulting from the lack of one or more vitamins.

Dentine. A substance that is relatively softer than enamel, forming the bulk of a tooth.

Depressant. A drug having an anaesthetic effect on the nervous system.

Diaphragm. A muscular partition separating the thorax from the abdomen. Also, a contraceptive device used as a barrier at the cervix.

Diastole. Part of the cycle of the heart during which the ventricles relax and receive blood from the atria.

Diffusion. The spreading out of molecules in a given space from a region of greater concentration to one of lesser concentration.

Digestion. The process during which foods are chemically simplified and made soluble for absorption.

Diploid. Term used to indicate a cell or an organism that contains a full set of homologous pairs of chromosomes.

DNA (deoxyribonucleic acid). A giant molecule in the shape of a double helix, consisting of alternating units of nucleotides, composed of deoxyribose sugar, phosphates, and nitrogen bases.

Dominance. The principle first observed by Mendel, that one gene may prevent the expression of an allele.

Duodenum. The region of the small intestine immediately following the stomach.

E

Ecosystem. A unit of the biosphere in which living and non living things interact, and in which materials are recycled.

Egestion. Elimination of insoluble, non-digested waste.

Embryo. An early stage in a developing organism.

Enamel. The hard covering of the crown of a tooth.

Endocrine gland. A ductless gland that secretes hormones directly into the blood stream.

Endoskeleton. Internal framework of vertebrates made of bone and/or cartilage.

Enzyme. A protein that acts as a catalyst.

Epithelium. A tissue that covers various organs and the body surface.

Epiglottis. A cartilaginous flap at the upper end of the trachea.

Eustachian tube. A tube connecting the throat with the middle ear.

Evolution. The slow process of change by which organisms have acquired their distinguishing characteristics.

Excretion. The process by which metabolic wastes are removed from cells and the body.

Exoskeleton. The hard outer covering of certain animals.

Expiration. The discharge of air from the lungs.

Extensor. A muscle that straightens a joint.

F

Faeces. Intestinal solid waste material.

Fallopian tube. See oviduct.

Fallout. Radioactive particles that settle on the Earth's surface from the atmosphere.

Fermentation. Glucose oxidation that is anaerobic and in which lactic acid is formed in muscle and ethanol by plants.

Fertilisation. The union of two gametes.

Fetus. Mammalian embryo after the main body features have formed.

Fibrin. A substance formed during blood clotting by the reaction between thrombin and fibrinogen.

Fibrinogen. A blood protein present in plasma involved in clotting.

Flexor. A muscle that bends a joint.

Food. Any substance absorbed into the cells of the body that yields material for energy, growth and repair of tissue and regulation of the life processes, without harming the organism.

Food chain. The transfer of the sun's energy from producers to consumers as organisms feed on one another.

Food pyramid. A quantitative representation of a food chain, with the food producers forming the base and the top carnivore at the apex.

Food web. Complex food chains existing within an ecosystem.

Fovea. The most light-sensitive spot on the retina of the eye where cones are most abundant.

G

Gall bladder. A sac in which bile from the liver is stored and concentrated.

Gamete. A male or female reproductive cell.

Gastric fluid. Glandular secretions of the stomach.

Gene. That portion of a DNA molecule that is capable of replication and mutation and passes on a characteristic from parent to offspring.

Genetic code. That sequential arrangement of bases in the DNA molecule, which controls characteristics of an organism.

Genetics. The science of heredity.

Genotype. The hereditary make-up of an organism.

Glomerulus. The knot of capillaries within a Bowman's capsule of a kidney tubule.

Gonads. The male and female reproductive organs in which gametes and hormones are produced.

H

Haemoglobin. An iron-containing protein, giving red blood cells their red colour and which carries oxygen around the body.

Glossary of biological terms used in the text

Haploid. A term used to indicate a cell, such as a gamete, that contains only one chromosome of each homologous pair.

Hepatic portal vein. A vessel carrying blood to the liver before the blood returns to the heart.

Heredity. The transmission of characteristics from parents to offspring.

Heterozygous. Refers to an organism in which the paired alleles for a particular characteristic are different.

Homeostasis. A steady state that an organism maintains by self-regulating adjustments.

Homologous chromosomes. A pair of chromosomes which are identical in form and in the way in which genes are arranged.

Homozygous. Refers to an organism in which the paired alleles for a particular characteristic are identical.

Hormone. The chemical secretion of a ductless gland producing a physiological effect.

Hypothalamus. Part of the brain that controls the pituitary gland.

I

Ileum. The longest region of the small intestine where digestion is completed.

Immunity. The ability of the body to resist disease by natural or artificial means.

Incisor. One of the cutting teeth in the front of both jaws in mammals.

Insecticide. A chemical that kills insects.

Inspiration. The intake of air into the lungs.

Interferon. A cellular chemical defence against a virus.

Involuntary muscle. One that cannot be controlled at will, like smooth muscle.

Iris. The muscular, coloured part of the eye behind the cornea, surrounding the pupil.

Islets of Langerhans. Groups of cells in the pancreas that secrete insulin.

J, K

Joint. The place at which two bones meet.

Kidney. An excretory organ that filters urea from the blood.

L

Lacteal. A lymph vessel that absorbs the products of fat digestion from the intestinal wall.

Larynx. The voice box.

Ligament. A tough strand of connective tissue that holds bones together at a joint.

Liver. The largest gland in the body, associated with metabolism of carbohydrate, protein and fat.

Lung. An organ for gaseous exchange during breathing.

Lymph. The clear liquid part of the blood that enters tissue spaces and lymph vessels.

M

Mammary glands. Those found in female mammals that secrete milk.

Medulla. In the kidney, the inner portion composed of pyramids, that in turn are composed of tubules; in the adrenal gland, the inner portion.

Medulla oblongata. The enlargement at the base of the brain called the brain stem. It controls the activities of internal organs.

Meiosis. The type of cell division in which, in the production of eggs and sperm, there is a reduction of chromosomes to the haploid number.

Meninges. Protective membranes surrounding the spinal cord and brain.

Menstruation. The periodic breakdown and discharge of the uterine lining that occurs after puberty in the absence of fertilisation.

Messenger RNA. The type of RNA that receives the code for a specific protein from

DNA in the nucleus and acts as a template for protein synthesis on the ribosome.

Metabolism. The sum of the chemical processes taking place in the body.

Mitosis. The division of chromosomes preceding the division of cytoplasm and leading to two identical cells from an original cell.

Molar. A large tooth only present in an adult's dentition and used for grinding; sometimes called a wisdom tooth.

Motor end plate. The end of an axon of a motor nerve in a muscle.

Mucus. A slimy lubricating and cleaning secretion from mucous glands.

Mutation. A change in genetic makeup resulting in a new characteristic that can be inherited.

Myosin. A form of thick filamentous protein that, together with actin filaments, composes a muscle fibril.

N

Natural selection. The result of survival of the fittest to breed in the struggle for existence among organisms possessing those characteristics that give them an advantage.

Nephron. One of the numerous excretory tubules in the kidney, including the Bowman's capsule and glomerulus.

Nerve cord. Part of the central nervous system extending from the brain along the dorsal side of the body.

Nerve impulse. An electrochemical stimulus causing change in a nerve fibre.

Neuron. A nerve cell body and its processes.

Nitrification. The action of a group of soil bacteria on ammonium compounds, producing nitrates.

Nitrogen cycle. A series of chemical reactions in which nitrogen compounds change form resulting in a stable concentration of nitrogen in the atmosphere and nitrate in the soil.

Nitrogen fixation. The process by which certain bacteria in the soil or in the roots of leguminous plants change free nitrogen into nitrogen compounds that plants can use.

Non-biodegradable. Polluting materials that are not decomposed by natural processes.

Nucleotide. A unit composed of deoxyribose sugar, a phosphate, and a base. Many such units make up DNA.

Nucleus. The part of the cell that contains chromosomes and which controls its activities.

O

Optic nerve. The nerve leading from the retina of the eye to the optic lobe of the brain.

Organ. Different tissues grouped together to perform a function or functions as a unit.

Organism. A complete and entire living thing.

Osmosis. The diffusion of water through a selectively permeable membrane from a region of greater concentration of water to a region of lesser concentration of water.

Ovary. The female reproductive organ.

Oviduct. One of a pair of tubes in a female through which eggs travel from the ovary and in which fertilisation occurs.

P

Pancreas. A gland located near the stomach and duodenum that produces digestive juices and hormones.

Pasteurisation. The process of killing and/or retarding the growth of bacteria in milk and alcoholic drinks by heating to a selected temperature so that the flavour is retained.

Pathogenic. Disease-causing.

Pectoral girdle. The framework of bones by which the forelimbs of vertebrates are supported.

Pelvic girdle. The framework of bones by which the hind limbs of vertebrates are supported.

Glossary of biological terms used in the text

Pericardium. The membrane surrounding the heart.

Periodontal membrane. The fibrous structure that anchors the root of a tooth in the jaw socket.

Peripheral nervous system. The nerves communicating with the central nervous system and other parts of the body.

Phenotype. The outward expression of genes.

Pituitary gland. A ductless gland composed of two lobes, located beneath the cerebrum.

Placenta. A large, thin membrane in the uterus that exchanges materials between the mother and the embryo by diffusion.

Plasma. The liquid part of blood.

Platelet. The smallest of the solid components of blood, which releases thrombokinase for clotting.

Pollution. The addition of impurities.

Population. A group of inter-breeding individuals in an ecosystem.

Prostate gland. A gland located near the upper end of the urethra in a male, helping to produce seminal fluid.

Protoplasm. Refers to the complex, constantly changing system of chemicals in a cell that establishes the living condition.

Puberty. The age at which the secondary sex characteristics appear.

Pulmonary. Pertaining to the lungs.

Pulse. Regular expansion of the artery walls caused by the beating of the heart.

Pupil. The opening in the front of the eyeball, the size of which is controlled by the iris.

Pyloric sphincter. A ring of muscle regulating the passage of semi-liquids from the stomach to the duodenum.

Q, R

Quarantine. Isolation of organisms to prevent the spread of infection.

Radioactive. Refers to an element that spontaneously gives off radiations.

Receptor. A cell or group of cells, that receives a stimulus.

Recessive. Refers to a gene or character that is masked when a dominant allele is present.

Rectum. The posterior portion of the large intestine.

Reduction division. See meiosis.

Reflex action. A nervous reaction in which a stimulus causes the passage of a sensory nerve impulse to the spinal cord or brain, from which a motor impulse is transmitted to a muscle or a gland.

Renal. Relating to the kidney.

Replication. Self-duplication, or the process whereby a DNA molecule makes an exact duplicate of itself.

Reproduction. The process through which organisms produce offspring.

Respiration. The release of energy from glucose in every living cell.

Response. The reaction to a stimulus.

Retina. The inner layer of the eyeball, formed from the expanded end of the optic nerve.

Rh factor. Any one of six proteins found on the surface of the red blood cells of most people.

RNA (ribonucleic acid). A nucleic acid in which the sugar is ribose. A product of DNA, it controls protein synthesis.

Rod. A type of cell in the retina of the eye that responds to shades of light and dark but not to colours.

S

Saliva. A fluid secreted into the mouth by the salivary glands containing the enzyme amylase.

Salivary gland. A group of secretory cells producing saliva.

Sclerotic layer (sclera). The outer layer of the wall of the eyeball.

Scrotum. The pouch outside the abdomen that contains the testes.

Sedative. An agent that depresses body activities.

Selectively permeable membrane. One that lets some substances pass through more readily than others depending on their molecular size.

Semen. Fertilising fluid consisting of sperms and fluids from the seminal vesicle, prostate gland, and Cowper's gland.

Semicircular canals. The three curved passages in the inner ear that are associated with balance.

Semilunar valves. Cup-shaped valves at the base of the aorta and the pulmonary artery that prevent back flow of blood into the ventricles.

Seminal vesicles. Structures that store sperm.

Seminiferous tubules. A mass of coiled tubes in which sperms are formed in the testes.

Sensory neurons. Those that carry impulses from a receptor to the central nervous system.

Serum. Plasma without clotting factors.

Sex chromosomes. The two kinds of chromosomes (X and Y) that determine the gender of a person.

Sex-linked character. A recessive charcter carried on the X chromosome.

Small intestine. The digestive tube, about seven metres long that begins with the duodenum and ends at the colon.

Smog. Combination of smoke and fog.

Smooth muscle. That which is involuntary and is found in the walls of the intestine, stomach and arteries.

Sperm. Short for spermatozoon, the male reproductive cell.

Sphincter muscle. A ring of smooth muscle that closes a tube.

Spinal cord. The main dorsal nerve of the central nervous system.

Spinal nerves. Large nerves connecting the spinal cord with the organs of the body.

Stimulant. An agent that increases body activity.

Stomach. An organ that receives ingested food, prepares it for digestion, and begins protein digestion.

Synapse. The space between nerve endings.

Synovial fluid. A secretion that lubricates joints.

Systole. Part of the cycle of the heart during which ventricles contract and force blood into arteries.

T

Taste buds. Flask-shaped structures in the tongue containing nerve endings that are stimulated by chemicals.

Tendon. A strong band of connective tissue which connects a muscle to a bone.

Testes. The male reproductive organs which produce sperms.

Thorax. The middle region of the body between the neck and the abdomen.

Thrombin. A substance formed in blood clotting as a result of the reaction of prothrombin, thrombokinase, and calcium ions.

Thrombokinase. A substance essential to blood clotting formed by breakdown of platelets.

Thyroid. The ductless gland, located in the neck that regulates metabolism.

Tissue fluid. That which bathes cells of the body. It is called lymph when in vessels.

Trachea. The windpipe taking air to the bronchi.

Transfer RNA. A form of RNA which delivers amino acids to the template formed by messenger RNA on the ribosomes.

Tympanic membrane. The ear drum.

U

Umbilical cord. The link between the embryo and the placenta.

Glossary of biological terms used in the text

Urea. A nitrogenous waste substance made in the liver from excess amino acids.

Ureter. A tube leading from the kidney to the bladder.

Urethra. A tube leading from the bladder to an external opening of the body.

Urine. The liquid waste made in the kidney and stored in the bladder, consisting mainly of water and urea.

Uterus. The organ in which developing embryos are nourished and protected until birth.

V

Vaccination. Method of producing immunity by inoculating with a vaccine.

Vaccine. A substance used to produce immunity.

Vagina. Cavity in the female immediately outside and surrounding the cervix of the uterus.

Vas deferens. Tubes that transport sperms from the testes.

Vector. A carrier of disease-causing organisms.

Ventricle. A muscular chamber of the heart for pumping blood.

Vertebra. A backbone.

Vertebrate. An animal with a backbone.

Viruses. Particles that are non-cellular and have no nucleus, no cytoplasm. They cannot reproduce unless they are inside living cells.

Vitamin. An organic substance that helps enzymes work in the body.

Vitreous humor. A transparent jelly-like substance that fills the interior of the eyeball and maintains its shape.

Vocal cords. Those structures within the larynx that vibrate to produce sound.

Voluntary muscle. Striated muscle that can be controlled at will.

W, X, Y, Z

White blood cells. Colourless, nucleated blood cells for defence.

X chromosome. A sex chromosome present singly in males and as a pair in females.

Y chromosome. A sex chromosome found only in males.

Zygote. The product of fertilisation.

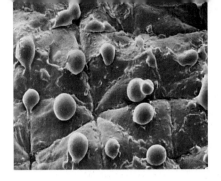

Index